Nomads in Central Asia

Animal husbandry
and culture
in transition
(19th - 20th century)

Carel van Leeuwen
Tatjana Emeljanenko
Larisa Popova

Royal Tropical Institute - The Netherlands

Nomads in Central Asia. Animal husbandry and culture in transition (19th - 20th century) accompanies the exhibition «Nomaden in Centraal-Azië», on show in the Tropenmuseum in Amsterdam from 1 December 1994 to 1 August 1995.

Address: Royal Tropical Institute (KIT)
Mauritskade 63
1092 AD Amsterdam
The Netherlands
(31) (20) 5688 711

Authors:
F.C. van Leeuwen, curator, Islamic culture, Tropenmuseum - Amsterdam
T.G. Emeljanenko, curator, The Russian Ethnographical Museum - St. Petersburg
L.O. Popova, curator, The Russian Ethnographical Museum - St. Petersburg
Photography: Jaap de Jonge, curator, Audio-visual programs, Tropenmuseum - Amsterdam

Cover photo: Kazakh shepherd in the steppes of Uzbekistan, 1993.
Reverse of cover: Detail of a woven strip, used to fasten the wooden frame of a yurta. Turkmen; height 24 cm. (Collection: Russian Ethnographical Museum - St. Petersburg)

Table of contents

Preface

The inhabitants of the five Central Asian republics which until 1991 were part of the Soviet Union, together form an amalgam of peoples whose histories reach far back. Some peoples arrived and settled long ago, as is the case with the Tajiks in Tajikistan. Others, like the Turkmen, the Kazakhs, the Kyrghyz and the Uzbeks are Turks, and arrived in the more recent past. The Uzbeks came from the lower Wolga area and settled from the 15th century onwards in what was then called Transoxania. This area comprised (the surroundings) of Bukhara Khiva and Kokand. Together with the Karakalpaks, living in northwest Uzbekistan south of the Aral Sea, they are now incorporated into the republic of Uzbekistan. Turkmen tribes settled in the 18th century, in what is now called Turkmenistan. The Kazakhs and the Kyrghyz are in fact one people. Soviet rule spread them over two republics: the Kazakhs in the steppes in the north and the Kyrghyz in the mountains bordered by the Alatau range in the south. These five republics represent what is presently called Central Asia, and before the Russian Revolution of 1917, Turkestan, land of the Turks. They cover an immense, sparsely populated, often empty landscape. The inhabitants of the now independent states of Kazakhstan, Uzbekistan, Kyrgyzstan, Turkmenistan all speak a Turkish language, whereas in Tajikistan an Iranian language is spoken. Russian is the lingua franca.

This book, accompanying an exhibition on material culture of livestock-breeders in Central Asia in the Amsterdam Tropenmuseum, describes the history of animal husbandry of some of the peoples in these five republics. Livestock-breeders in Central Asia used to be nomads, with large herds of horses and sheep. Relatively small distances were covered with the sheep, wandering from low-lying areas to the mountain plateaux in summer and returning in fall to the winter residences. Only in middle and northern Kazahkstan and parts of Uzbekistan and Turkmenistan were much longer distances covered to and from the traditional pasture lands. The arts and crafts of these livestock-breeders have been closely related to their nomadic life style, as can be witnessed by traditional felt making, leather working, weaving and textiles, weaponry, dolls, ritual objects, clothes, yurta furnishing and objects referring to animals. However, lives were changed dramatically after the advance of tsarist Russia in the 19th century, and the consequences of these changes for nomadic (material) culture were just as dramatic.

The penetration by Russia in tsarist times through the merchants and the army, followed by 'western' (Russian, Baltic, Ukrainian, etc.) colonists profoundly changed Turkestan. Pasture lands were converted into farmlands. European influence made itself felt as well. Various rebellions, especially during the First World War, against tsarist mobilization policies led to the death or flight of thousands of peoples. The venue of bolshevism after 1917 disrupted even more: the collectivization in the twenties and thirties took a more than heavy toll on people and animals. As a result of tsarist and bolshevik politics and policies, traditional nomadism was annihilated. This had its repercussions on the material culture of the peoples in the southeastern corner of what would become the Soviet Union. During the 20th century, the beauty, vitality and brightness of that region disappeared from public life, to be replaced by dull uniformity.

What traces did nomadic culture leave in Central Asia? Travelling through the republics of Kazakhstan, Uzbekistan and Kyrgyzstan in the autumn of 1993 and the spring of 1994, this question proved not easy to answer. Animal husbandry is still dominant in vast areas of Central Asia and its organization reflects ancient social relationships. Daily life, however, is profoundly changed. These historical developments are discussed here, illustrated with various objects from the Russian Ethnographic Museum in St Petersburg. Not much has been written about nomadic culture in this area, a great deal in Russian, or other languages which are not easily accessible in western Europe. In the following chapters, Larisa Popova and Tatjana Emeljanenko describe animal husbandry and the life of herdsmen and their families in the 19th and 20th centuries, based on this body of literature and museum collections.

For an analysis of nomadic animal husbandry in Central Asia and changes in nomadic culture, travelling through the region can complement this history stored in museum collections and books. Much information in the first chapter of this study was obtained by discussing past and contemporary animal husbandry with present-day livestock-breeders and others involved. Carel van Leeuwen interviewed different herdsmen and kolkhoz members. Information was obtained as well from Ms. G. Kunaiva, a staff member of the National Museum in Almaty, Kazakhstan; the various staff members of the privatized sections of the Ministry of Agriculture, Department of Animal

Husbandry, and V. Schimpf, the former director of a kolkhoz near Karaganda, Kazakhstan. From these interviews an impression arose of a life in difficult climatic conditions, hot in summer, cold to extremely cold in winter. The vastness of the area and the good soil quality still leaves the Central Asian livestock-breeders possibilities to make their living out of animals. This remained unchanged, notwithstanding the Russian advance in the 19th century and the measures and steps taken by Soviet rulers from 1925 onwards. However, the landscape has changed. The advance of agriculture strongly influenced life in Kazakhstan, Uzbekistan, Turkmenistan and Tajikistan. Cotton and fruits dominate the valleys and basins of the rivers where formerly the animals grazed.

Kazakhstan now produces sugar beet, potatoes, wheat and other agricultural products. But the heart of many farmers is still bound to their sheep, horses, and camels.

RUSSIA
KAZAKHSTAN
Caspian Sea
Lake Aral
Aqmola (Tselinograd)
Syr Darya
Amu Darya
KARAKALPAKS
UZBEKISTAN
TURKMENISTAN
Lake Balkhash
Almaty
Bishkek
KYRGYZSTAN
Toshkent
Samarkand
Bukhoro (Bukhara)
Ashkhabad
Dushanbe
TAJIKISTAN
CHINA
IRAN
AFGHANISTAN
0 250 500 km
0 300 miles
Ethnolinguistic groups of Central Asia (1989)
Peoples speaking aTurkic language
Kazakhs
Kyrgyz
Uzbeks
Turkmen
Karakalpaks
Uygurs
Bashkirs, Tatars
Peoples speaking an Iranian language
Tajiks and peoples of Pamir
Peoples speaking a Slavic language
Russians
Ukrainians
Poles
Mixed population of Russians/Ukrainians and Kazakhs
Others
Germans
Koreans
Meskhets
Sparsely populated area

1 Animal husbandry in Central Asia: rooted in the past

Carel van Leeuwen

Before the Mongol invasions of the 12th - 14th centuries, the Tajiks formed the majority of the peoples in Central Asia. They lived in the basins of the two rivers Amu Darya and Syr Darya, with Samarkand and Bukhara [today called Bukhoro] as the most important towns. From the 10th century onwards, civilization flourished. Many remnants of the glorious past of Bukhara, and of Samarkand, can still be seen and admired, although according to contemporary writers like Ibn Battuta the Mongol hordes left the whole oasis in ruins. After the upheaval of these Mongol invasions, the ethnic map of this area was completely altered. Many Tajiks were forced by the Uzbeks to retreat to the foothills of the Pamir. Under the Uzbek Shaibani dynasty a symbiosis gradually arose between the Turkish nomads and the sedentary Tajiks, who still formed the bulk of the population in the area bordered by the Amu Darya and the Syr Darya.

Through Bukhara, where Islam had entered in the 8th century, Islam spread through the area between the Caspian Sea and the Russian empire from the 16th century onwards. Muslim merchants from the Arab world, Iran and Turkey on the one side and from Kazan in the north on the other, served as go-betweens. Islam created a certain sense of 'belonging together', enhanced by the absence of language barriers, in the whole of Central Asia. Because of the trade, economic relations between Russia and Central Asia were intensified.

Centuries of change

At the beginning of the 18th century a new dynasty seized power in Bukhara. This dynasty would rule until the Russian Revolution in 1920. Turkestan, as it was called, was divided into three units, the Emirate of Bukhara and the Khanates of Kokand and Khiva. The relative backwardness of these powers in the 19th century, as compared to the earlier flourishing civilization, was mainly due to the loss of preponderance in overland trade relations between both east and west and north and south. Travel and trade overseas was safer and quicker. Also, mutual relations between the emirate and the khanates in the first half of the 19th century were strained. When these relations improved during the second half of the century, this had a positive effect on the economic situation. Agricultural production increased. The state (56%), the Emir (12%) and religious orders (24%) owned the bigger part of the fields which were worked with sharecroppers, serfs and slaves. Only 8% was owned by others (Carrère d'Encausse, 1988:9). According to Vambery, quoted by Rypka (1968:512), 20 000 slaves were active in the emirate in the second half of the 19th century, most of them captured during raids in northern Iran. The economic growth, following the relative peace, resulted in the increase of commercial contacts with the Russians.

In Russia capitalism developed from the middle of the 19th century onwards, as was the case in western Europe. Russia looked for markets for its growing production and at the same time for raw materials like cotton. Therefore, Russian interest in getting more influence in Turkestan increased. Whenever this suited the Russians, the emir of Bukhara was assisted in suppressing rebellions in various parts of the emirate. Kagan, a few kilometres east from Bukhara, became the headquarters of the tsarist political officer. Although the emirate and khanates remained independent, Turkestan gradually became dominated by Russia.

Russian foreign policies were dictated by the tsarist efforts not to strain relations with the British empire through waging a war over India. The emirate was the bone of contention between the Russian empire and the British empire. Both vied to incorporate the area in their sphere of influence for geopolitical reasons: the Russians in order to penetrate the communication lines between England and British India, whereas England sought to protect these and keep the Russians at a distance. However, in 1868 a small but well-equipped Russian army defeated the emir's army weakened by crushing rebellions in Kokand. The emir received no support from the population, which was heavily burdened by taxes. Bukhara formally remained independent, but Russian tutelage was strong. Until the beginning of the 20th century the emirate was ruled in a traditional feudal way. Civil servants received no salary and there was no yearly budget. Slavery existed in all parts of the emirate and was only abolished under Russian pressure. However, under Russian influence, the feudal system gradually broke down.

Russian rule did not differ from colonial practices elsewhere in the world. An influx of Russian colonists took place, the markets were flooded with Western goods outdoing the locally made artisanal goods. Raw materials, like cotton and minerals, went northwest. Tsarist Russia brought an end to the countless local wars in Central Asia, slavery was abolished and a Western-style educational system developed. Illiteracy had been nearly one hundred percent, the only schools being confined to mosques where religious teaching was primordial. The Russians strengthened agricultural development and laid the foundations for industrialization.

In the 20th century the first industries were established in Central Asia, and immigration of westerners took place. Before the outbreak of the First World War fifty companies had started operations. Banks had opened branches. Cotton production was promoted which, however, led to shortages in the food production. The construction of the trans-Caspian railway accelerated integration. The first Russian Revolution from 1905-1907 did not affect the local population but the Russian railroad workers were the first to revolt. The army crushed the rebellion. In 1920 in the aftermath of the October Revolution the emir of Bukhara was toppled.

The sociopolitical situation in Turkestan remained undecided from 1917 onwards till the beginning of the thirties. The local population had a difficult choice between the old and the new regime (Lemercier-Quelquejay, 1985:36). In this period thousands fled and thousands died, including the pre-revolutionary local intelligentsia which was wiped out in bloody purges. Anti-Soviet rebels, among them the then well-known Basmachis held out till the beginning of the thirties. From then on, integration of Turkestan in the Soviet Union gained momentum. The wounds inflicted in the preceding war are not yet forgotten. Nor is forgotten the subdivision of Turkestan into various republics, ordered by Stalin in the twenties. The frontiers of these republics were chosen in such a way that every republic contained large minorities of neighbouring republics (see especially Bennigsen and Wimbush, 1985). The 'divide and rule' principle, applied all over the world by colonial powers, was applied here as well. For instance, the Tajiks still resent the inclusion of Samarkand, a town where Tajiks make up more than half of the population, in the republic of Uzbekistan.

Transformation processes had started after the slow but persistent penetration of Western colonists in the 19th century. The Soviets, with their continued promotion of modern techniques, trade and infrastructure, accelerated these processes. Nomadic animal husbandry increasingly disappeared.

The trans-Caspian railway was built to encourage Russian exports and promote Russian expansion. (Photo collection, Central State Museum of Kazakhstan, Almaty)

One of the oldest buildings in Almaty is the Russian orthodox Zenkov cathedral built in 1903/1904.

Nature, rituals and nomadic religious life

Islam and the encounter with shamanism

Between 705 and 715 AD (Carrère d'Encausse, 1988:1) Islam was introduced in Bukhara. This religion would gradually become predominant, not only in Bukhara, but in the whole of Transoxania, especially under the sedentary population in Turkmenistan, Tajikistan and Uzbekistan. From the 8th till the 16th century, Central Asia was at the centre of the Islamic world. The fame of Samarkand and Bukhara – where under the patronage of successive rulers science and technology flourished – was known all over the world. Through these cities the North-South trade (the fur route) and East-West trade (the silk road) brought wealth, and new ideas were spread. From the 8th century onwards Islam intermittently and gradually conquered Central Asia. Through the military activities and conquests by Persians, Caucasians, Arabs and Turks, Islam became firmly entrenched in the oasis.

Only after the turn of the 19th century, with the growing incorporation of the nomads in sedentary social structures through Russian influence, did Islam gain some foothold in the world of the Central Asian steppe nomads. Before, these steppe nomads practised long-distance nomadism. Only during short periods did they stay with their herds in the same place, moving, when necessary, on to other traditionally demarcated pastures. Only at annual fairs might they meet Islamic merchants and missionaries. These meetings were too short for conversion.

The immense area between Russia and China – bordered in the south by the mountain-ranges of the Himalayas, in the west by the Caspian Sea, in the east by the Tien Shan mountains – experienced various religious influences. Among them were Buddhism and Lamaism, Islam, Manichaeism[1] and Nestorian Christianity[2]. Among nomads in Kyrgyzstan and Kazakhstan, shamanism was influential. According to Czaplicka (1973:29, confirmed by Bennigsen, 1967:180 and by Krist, 1992:45, 97) this was the case even in the 20th century.

Shamanism and nomadism

Characteristic of shamanism is the idea of the existence of two groups or types of gods: those who bless and give prosperity, and the gods that bring evil.

Good gods operate above the earth. The evil gods reside under the earth. Their realms touch each other on earth. The shaman will, during ceremonies, evoke the protection of the benevolent gods and the deterrence of the destructive forces of the malevolent gods.

Shamans are specialists. They are able to contact the spiritual world around us and to influence this world. The shaman is the intermediary between this world and the spiritual world.

Shamanism can be found among hunters and gatherers and among nomads. Their wandering way of life, close to nature, partly shaped shamanism. Shamanist perceptions of the world are strongly determined by the mysterious and violent surroundings. These surroundings, this real world, is also seen as personified. Stones, plants, animals, forces of nature possess a soul and each of them has its own identity. The cosmos too – as a whole as well as its various parts – has experiences, has its own will, feelings and its own consciousness. And all these phenomena influence each other. In ancient times they could live together and communicate. This paradise on earth ended with an incident equivalent to the Fall. Myths will refer to this incident. Only shamans who entered and know heaven can travel back to the mythical prehistoric times through their experiences and knowledge. They are able to convey the lessons learned and the messages from the supernatural world to the members of their group or tribe.

The shaman is active in the group from which he originates and to whom he extends his services. He is a mystic, dancer, medical doctor, poet, mythologist, magician and priest. He accompanies the dead on their way to the underworld, he tries to accompany the sick back to the real world, to wrest happiness from the souls of the animals for the benefit of the members of the group or tribe, to see into the future whereby he discusses the fate of the people with the supernatural. He tries to recover lost objects, to give advice and offer protection against the unknown and unseen. His most important task, however, is to influence as positively as possible the well-being of the members of his group. For the group the shaman is the religious institution providing order in the world and the community. Sometimes shamans designate one of their children as their successor. The position of shaman, however, is not hereditary. Shamans may be selected by supernatural powers or otherwise a physical sign or peculiarity is instrumental. Orphans especially may be selected as trainees and successors (Eliade, n.d.:30).

The group or tribe has to recognize the shaman. After a period of practical training he devotes himself to satisfying his group through visions which he communicates. To reach his objectives he uses dance, poems, singing, sometimes hallucinogens like special types of fungi (Rosenbohm, 1991:intro), together with musicals instruments, amulets, special clothes and a headgear. On his drum are signs and symbols with mythical and cosmological contents. Drums are vehicles through which a shaman can reach the other world. Parts of his clothing refer to this ability to come into contact with others.

Shamanism often coexists with other forms of magic and merges with other religions. When, for instance, a shaman visits sick people, he will invoke the name of Allah and other famous Islamic saints before starting his curing activity. Evil spirits and djinns will be threatened and driven away. After the seance – during which the shaman will often be in a trance – he will again invoke Allah.

Notwithstanding strong pressure by the communist authorities, remnants of shamanism still exist among the Kazakhs and Kyrghyz. The position of the bugsa, shaman, is hereditary there (Castagné, 1930:10). Traces of the ambivalent relations with other worlds can also be found on the traditional jewellery of the Turkmen and the Kazakhs, with their references to forebears, the supernatural and the underworld (Schletzer, 1983: e.g. 40, 43, 60, 160). The Turkmen, however, came much more under the influence of Islam than the nomadic Kazakhs and the Kyrghyz, living in the less accessible mountainous areas. Islamic leaders tried in vain to banish traditional Turkmen jewellery because of its reflection of religious views which conflicted with official Islam.

In the second half of May 1994, a Kazakh living in the town of Uzyn Agach, 50 km west of Almaty (Alma Ata), joined us when we left for Tambeltas ('place with drawings'), a holy place two hours northwest of the town. The place is watered by a small stream and consists of small hills, providing shelter against the climate both in winter and summer. It is the site where, in the past, repeated clashes took place between the Kazakhs and the Djungarians. According to tradition this place had sheltered Mani for a long time, during his journey eastwards. It is also said that today, a shaman sometimes retires there to prepare himself.
Part of the rocks is covered with drawings of men and animals, and also some human figures. Near the stream shrubs and bushes are growing. On their branches small strips of white cloth, nearly all made of cotton,

An inhabitant of Uzyn Agach asks blessings for his family by tying a small strip of cloth to one of the bushes in Tambeltas, a holy place three hour's drive west of Almaty, Kazakhstan. The prehistoric rocks at this place are decorated with engraved drawings. The drawings show people and animals like sheep, horses and camels.

Shaman treating a patient. (Photo collection, Central State Museum of Kazakhstan, Almaty)

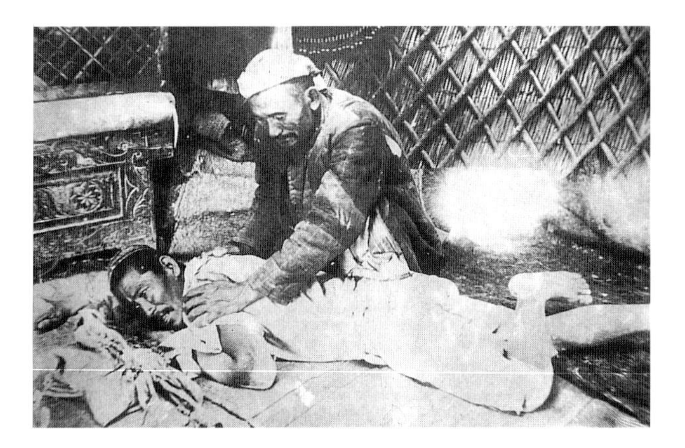

are attached. Our companion tore his handkerchief and tied a small strip to one of the bushes, asking blessings for his family. On festive days like nawruz or during a marriage feast, people will visit Tambeltas and perform rites which refer to shamanism or, as was said by local people, fetishism.

▬

Islam and sedentary life

Soviet rule has been trying to minimalize and control all religious activity, but Islam in Central Asia flourishes till this day. MacLean (1959:87) quotes Anthony Jenkinson, a merchant adventurer from England who visited Bukhara in 1558. Jenkinson describes Bukhara as a 'great citie' with 350 mosques and more than a hundred medressas. During the Soviet period from about 1920 till 1990/91, nearly all of them were torn down. Three mosques and one medressa, the Mir-i-Arab were left. The last sharia court was closed in 1928 (Bennigsen, 1985:10). And this happened not only in Bukhara. Nevertheless, in 1985 the Soviet Union had fifty million Muslims, living mainly in Caucasia and Central Asia. Nearly all of the 37 peoples in the former Soviet Union labelled Islamic, lived together with non-Islamic minorities from other parts of the country (Idem, 26), mostly Russians and Ukrainians. This had also become the situation in the Central Asian republics.

Soviet authorities recognized only the Shi'a and Sunni sects, by far the two most important Islamic groupings. They sponsored regional congresses of Muslims through directorates led by laymen and the clergy. The most important was the directorate in Tashkent. It was the only directorate with access to financial resources, enabling publications. Also, the only two medressas in Central Asia where future mullahs were trained and educated, were situated in Tashkent. All religious activities not sponsored by the directorate were banned. A very small group of some two or three thousand Islamic officials were employed in the different directorates. These concerned imams, muezzins, qadis and other personnel, serving 50 million believers.

Both Western and Soviet sources used to distinguish between formal 'cultural Islam' and 'religious Islam'. The first refers to the set of values, practices and rites which are commonly adhered to. The second to the group of active followers. This distinction is obsolete now, as is the distinction between the Islam controlled by the state through the official directorates and the 'shadow Islam' (Halbach, 1991:10 et seq.). When speaking about Muslims, they are not only those who actively participate in religious life, but also people who share the cultural identity and accept the social conditions prevalent in Islamic countries. A Muslim from the southeastern part of the former Soviet Union is someone whose outlook and way of life is determined by the centuries old influence of Islam. His political, social and religious values are shaped in this way, but also influenced by shamanism and nomadic culture.

Sufi brotherhoods

Important elements of the more or less hidden Islam of this century were, and still are, the orders and the congregations of Muslims in the quarters of cities, towns and villages. The Sufi brotherhoods, *turuq* (sing. *tariqa*: path) still have many followers in Central Asia, both for religious and social reasons. These orders are hierarchically organized, they indicate the path to follow in order to reach God. During the Soviet period these orders attracted more followers than the officially organized meetings. People assisted in rites near the tombs of holy men and mystics, and came to holy places and holy trees. In the cities and towns the informal leaders of the various quarters assisted in the more social rituals concerning the life cycle, such as marriage ceremonies and relations between the families of the bride and groom. This informal leadership also gives assistance to those with many children who are financially not capable of marrying off all their children. The maintenance of good relations between the families and between couples in order to prevent divorce is also their responsibility. Burial rites and the celebration of special religious festive days like the *kurban bayram*, the *uraza bayram* and the *mawlid el Nabi* are observed by almost everybody. The *nawruz*, celebrated especially in the southern parts of Central Asia, was forbidden under Soviet rule. Now it is seen as a national day (Bennigsen, 1985:23).

Islam today

Central Asia has a considerable, even strong population growth, compared to other parts of the former Soviet Union. However, people strongly adhere to practices of traditional culture, the culture which stems from centuries-old customs. Seventy years of communist control could not erase this. In 1986 Soviet authorities condemned illegal religious practices and organizations, the flirting of party officials with Islam and ritual practices of Islam, and against

In a mahalle in Uzyn Agach, Kazakhstan, we were in October 1993 invited to join a commemoration. One year before, the father of one of the elders of an extended family of livestock-breeders had died. About sixty neighbours jointly assisted in a meal of slaughtered sheep and a slaughtered horse. An itinerant mullah, coming from Karaganda, presided over a short ceremony, reciting verses from the Quran (on the picture sitting in the middle of the front row). The eldest of six brothers (last row, to the right) was the host. His wife and three 'resident female attendants' – in short his four wives – and their daughters had prepared the meal. Dachšlejger's remark, 'Elemente der Polygynie (...) verloren für immer und unwiderruflich ihre Daseinsberechtigung' (1981:121) could not be more wrong.

the growing support of pan-Islamism (Halbach, 1991:27). The influence of (the aftermath of) the Afghan War was certainly felt in this matter.

Today, most of the Muslims in Central Asia (in 1991 more than 37.5 million out of 50 million) belong to the Sunni sect. The second largest, much smaller, group are the Shi'a, concentrated in the towns and cities like Samarkand, Ashkhabad, Mary (Merw) and Bukhara. They are descendants of Persians, captured during raids by Uzbeks and Karakalpaks before the rise of communism, or are members of the formerly powerful and rich merchant families. In the eastern part of Tajikistan, Gorno Badakhshan, there live about 100 000 Ismailis, followers of the Agha Khan.

Islam – in Tajikistan more than in other republics – seems to be used as a vehicle to strengthen the creation of a national identity. Together with political parties of different nature and ethnic alliances, this religion is one of the factors which will determine the future of the former Soviet societies in Central Asia. Islam is, however, no monolithic entity, and daily life is influenced as well by ethnic, political, socioeconomic and ecological factors.

The Russian conquest of Central Asia (1725-1914)

Almaty today knows a relative freedom and tolerance to promote religion. Religious books and pamphlets, for instance, both Islamic or Russian orthodox, can be obtained.
The construction of a new mosque was well advanced, when we visited Almaty in 1993. The existing one-storey mosque is a rather modest and unpretentious building. Still, the authorities are more or less reluctant in promoting religiosity. Real assistance in the construction of mosques is not given. Buschkow (1993:27) and Gressler (1993:30) both mention the financial assistance of Saudi Arabia in this matter. Buschkow mentions, however, the opinion of members of a fundamentalist party who describe the Saudi Arabian assistance as pitiful. Russians have repeatedly requested the re-opening of the Zenkov cathedral in central Almaty. This wooden, main orthodox cathedral from 1903/1904, was still closed in May 1994, although officially the re-opening was announced. This is just one example of social conflicts, fought out over religious institutions; in this case between Kazakhs and Russians.
To this example a conflict between Germans and Kazakhs can be added. At present, schoolchildren are forced by the school authorities to learn Islamic prayers and rituals. No attention is given to the fact that sometimes a (large) part of the children are from Lutheran parents (Gressler, 1993:19 and 30-31).
In Gazgelin, west of Almaty, the complaint was heard that the government did nothing to help the construction of a mosque. No construction materials were forthcoming. It was also worth noting that the local Kazakh population did not show any enthusiasm either. Of course, the enthusiasm for Islam can not be measured by the speed with which mosques are presently built. The shortages of building materials must be reckoned with. However, Islam's role in Central Asian society should not be overestimated. Even if Islam is present, it seems to be used to bolster (a part of) the national identity (Halbach, 1991:15 et seq.). To the trained ears of those who travel extensively in Northern Africa and the Near and Middle East, a conspicuous element of society is missing in Central Asia: the call to prayer. And this cannot only be caused by the relatively high cost of the purchase of sound equipment. Besides, the only time an individual (morning) prayer was heard during a six-week trip in 1993, was in the house of a Kyrghyz near Isfara in Uzbekistan.

Nomadic social relations and national boundaries

Tsarist and later Soviet influence and power in the 19th and 20th centuries brought the local population in the southeastern area of the former Soviet Union into close contact with westerners and thus unbelievers. In the first half of the 19th century, these contacts had not been manyfold. Both in areas where the population was sedentary, as well as in those where animal husbandry dominated, traditions ruled the social structure. Each individual belonged to an extended family along patriarchal lines; the family was part of a clan and tribe.

However, distinctions between the sedentary and the nomadic population were strong. In Tajikistan, Uzbekistan and partly Turkmenistan where Islam has, for centuries constituted the decisive factor, the sharia determined family and hereditary law. In Kazakhstan, Kyrgyzstan, Karakalpakstan and partly Turkmenistan, customary law, *adat*, was followed. For nomads the feeling of belonging to a particular group/family/clan was stronger than the unity of Islam, whereas the sedentary people felt themselves in the first place Muslim, and next a member of a specific town or village community. For them ethnic affiliations played a less important role or none at all, if compared to the nomadic peoples.

The contacts with the Russians strengthened the feeling of belonging to one community. But still, the common basis of the community was weak, historical distinctions were stronger. Kazakhstan, for instance, was socially made up of three hordes which were constituted by clans and tribes. Turkmen society was stratified and divided by the nobility, 'white bone', on the one hand, and the common people, 'black bone', on the other. The feeling of belonging to a nation was in fact prevented by the feeling of belonging to the family and to higher social divisions like clan and tribe. Kinship ties between families and clans confirmed the social and moral traditions in the whole of Central Asia. Only in the Turkmen society can an exception be found.

Due to sedentarisation in the 19th and 20th centuries, ancient clan and tribal traditions lost their impact on many nomadic clans. Besides, at the turn of the century, in the small circle of the local intelligentsia, interest was expressed in 'national' feelings, in a nation. This notion of nation referred to the amalgam of all the peoples (Turkmen, Tajiks, Uzbeks, Hazaras, Arabs, Persians, Afghans, etc.) inhabiting the area concerned. However, distinctions continued to exist. We will give a brief survey of them in the five new republics.

Kazakhstan Among the different peoples of Central Asia the Kazakhs have a special place. Before the October Revolution they were mostly nomads. Only quite near the mountains in the south, the area of the Seven Rivers (Semirech'e), were some people engaged in agriculture. The communist period did not destroy their tribal affiliations. The overall subdivision of society into three hordes, the *zhuz*, is still strongly felt. Each Kazakh belongs to a tribe of one of the three hordes: the Great, Middle and Lesser hordes, in southeast/east, middle/central/north and west Kazakhstan respectively. The *zhuz* are subdivided into clans and subclans. The present kolkhozes, consisting of one or more villages/settlements, often correspond to subclans.

The sedentarisation of the Kazakh people started in the 1920s, some years after the overthrow of tsarist rule. Presently, this sedentarisation period and the ensuing collectivization, is called the 'War against the people' by Kazakh opposition groups (Grobe-Hagel, 1992:31-32). The brutal force with which the authorities imposed their changes resulted, within three years, in the loss through death or flight of one quarter (1 230 000 persons) of the Kazakh population. Yet, nomadism was never completely wiped out. Even today, according to the director of a kolkhoz near Karaganda, not a few Kazakhs prefer to live in yurtas instead of the kolkhoz houses. The houses are used as stables and/or for storage.

Before the revolution, livestock had belonged to families which were a part of larger social organizations like the clan and tribe. Sheep were at the basis of life. During the Soviet period private ownership of land and livestock was abolished. The results were not only economically disastrous. The 'War against the people' had also taken a heavy toll of the animals. In Kazakhstan the total number of sheep declined from 36.3 million in 1929 to 3.3 million in 1933 (Grobe-Hagel, op. cit.; see also Bennigsen, 1985:70). One of our interlocutors in Almaty mentioned even higher numbers, stating that during escape journeys eastward, thousands of people and animals died. Another reason for the diminution of the number of sheep, was the instruction by Soviet authorities to shear sheep in wintertime, a measure probably dictated by a lack of wool. Sheep not protected by wool in temperatures of minus 20 to 30°C, however, do not have a big chance to survive in a country where a saying goes: *when you enter the winter with a potbelly you'll come out in spring lean and thin; when you start the winter thin you will not see the next spring.*

Kazakhstan can be divided into roughly two regions. In the first one, south and central, the majority of the population is Kazakh, whereas in the north and west, they are a minority. Various old customs are still adhered to in the areas where Kazakhs outnumber others, although authorities have tried to suppress some of them. When the Kazakh republic was founded the *kalym*, bride price, whatever its form, was prohibited by law. The freedom of women to choose their husband by free choice sought after by the Soviet authorities constituted the underlying reason. Other traditional institutions like 'betrothal in the cradle', marriage of unborn children, or levirate were abolished as well, although its effectuation took time. In the north and west of Kazakhstan, where the Kazakhs are a minority, traditions are disappearing fast.

Kyrgyzstan The relative isolation of mountainous Kyrgyzstan supported the preservation of the traditional social structure, especially in the rural areas. Groups of extended families, recognizing a common ancestor, basically form the nuclei of local communities. As in Kazakhstan and Uzbekistan, a kolkhoz consists of such extended family groups, in one or two, and sometimes three villages. The family groups are part of clans and tribes. Political leadership is rooted in these clans and tribes.[3]

Islam penetrated the region in the 17th and 18th centuries through the Fergana valley. Northern and eastern Kyrgyzstan are much less Islamised than the west and south. Those Kyrghyz who were converted to Islam mostly adhere to Sufi orders. In the Fergana valley, where the influence of different Sufi orders is strong, the most venerated shrines and holy places of Central Asia can be found.

Uzbekistan In neighbouring Uzbekistan the situation is different. The population belongs to various ethnic groups, although the Uzbeks constitute an overwhelming majority in the republic. Apart from recent immigrants from the former Soviet Union, the population can be subdivided into three major groups: those who settled in Uzbekistan before the 16th century; descendants from the Shaibanid Uzbek tribes, the last invaders from the (north)east; and the urban population, composed of ethnic groups with an old Iranian background.

The second group – descendants from the Shaibanid Uzbek tribes – lived a nomadic life, until the turnover of 1917. A fairly large number of tribes belonging to this group still feel strongly attached to tribal customs, and feel themselves related to Kazakhs, Kyrghyz, Karakalpaks, and others. Here as well, traditional rules concerning marriage, dating from before the revolution, largely prevail. Also people deliberately choose their partners from affiliated tribes. As is the case in other republics tribal affiliation and kolkhoz composition often coincide. Members from local families and tribes form the official leaders and representatives in important councils, in (before the turnover of the nineties) the communist party, as well as, more in general, in the national government.

In the urban centers, the mahalle is the social unit: a neighbourhood led by elders, with its own cemetery and house of prayer.

Tajikistan As the Tajiks have lived a sedentary life since time immemoral, tribal ties lost their importance completely. The largest social unit is the extended family. A village used to consist of one or more extended families. Leadership is retained by the elders and traditional customs (for instance the institution of the bride price, marriage at a young age, or levirate) still prevail.

Turkmenistan In Turkmenistan tribal clan affiliations are still strong. Before the Russian expansion in the 19th century, no confederation of tribes existed. Tribal loyalties and kinship ties have played an important role in society until this day. The Tekkes are the strongest Turkmen tribe; from their ranks the elite is selected. Each kolkhoz consists of a few extended families. Common ancestors play an important role in family and clan life; they strengthen social relations. Also endogamy within a clan is preferred. Traditional customs, as mentioned above, are adhered to.

One Sunday in 1994, we talked to V. Schimpf, the former director of a kolkhoz in Kazakhstan. He concluded that, although in his view society had changed, it was not as thoroughly restructured as the Soviets had intended. Indeed, the influx of westerners, and of displaced peoples from other parts of the Soviet Union had altered society. But the organization of life in the villages had stayed as before, the same families being influential, through their family ties, in the party and the state. And now, after the dissolution of the Soviet Union and the political formation of independent republics, this again would not change. And another interlocutor stated: 'Important people stay important and they keep their social network intact for the benefit of themselves and their kin'. These statements may be true for most parts of Central Asia.

Animal husbandry in Central Asia

From archaeological finds we know that Central Asia has, for a long time, known a symbiosis between nomads and the sedentary population in the rather densely populated basins of the Amu Darya and the Syr Darya, and at the foothills of the mountain ranges in the east of Tajikistan and Uzbekistan, or the south of Kazakhstan (Oxus, 1989). The Russian expansion of the last two centuries not only affected the sedentary population, but also changed nomadic practices. Through the influx, since tsarist times, of Byelorussians, Russians, Balts and Ukrainians, traditional pastures in Central Asia, especially in Kazakhstan and Uzbekistan, changed into farmlands. In the Soviet period more and more people were transferred, sometimes by brute force. This meant a steady decrease in grazing areas available for the local population. The collectivization that began in 1929 decreased this area even more.

In shock waves the nomads were forced to give up nomadism and to settle, enforced by the collectivization policy of the Soviet period. Animal husbandry had to be organized into sovkhozes and kolkhozes. At present, the overwhelming majority of the herds is managed by the animal husbandry sections of these kolkhozes and sovkhozes. Livestock breeding is not everywhere the same. In the steppes the herds circulate around the different farms of the kolkhoz. In winter the animals stay in winter stables. In spring, the relatively short summer and fall, they are led to the customary pastures. Near the mountains, however, traditional transhumance still takes place. The herds are brought from spring onwards to customary pastures on the high plateaux and returned in winter to the winter stables. Nomadism, in the sense of wandering through the steppes looking for food and, in winter, for shelter, does not exist anymore. Extensive animal husbandry practices is a better description of the present situation.

These changes, however, did not change the essence of animal husbandry and the climatological and environmental factors which ruled this before. To these we will turn now.

Central Asia and the Arab world

Many parallels and differences can be discerned, when comparing animal husbandry in Central Asia and the Arab world. At first sight, animal husbandry practices in both areas have similarities: large and small animals are kept; relations with the market are maintained through direct or indirect sale of animals to the inhabitants of villages and larger urban areas; livestock owners often travel over large distances; and customary law determines the boundaries of the clans' grazing areas.

Sheep, mostly of the Merino type are the core of the Central Asian herds, as they are in the Arab world. The availability of water is crucial for sheep. Only during very short periods in spring when persistent rains give sufficient moisture, can they do without the daily watering which is necessary during the rest of the year. The grazing range of sheep herds is therefore restricted.

Differences between Central Asian and Arab animal husbandry occur as well, for instance, as far as the animals are concerned. In Central Asia as in the Arab world, camels were used as pack animals. In the 19th century many camels still lived in Central Asia. They gave prestige to the owner and constituted a symbol of wealth. Communist rule suppressed symbols of wealth and camels gradually became replaced by trucks. However, camels eat especially thorny plants that the other animals despise, and which are harmful to horses and sheep. A side effect of modernity therefore was a deterioration in the quality of the steppes, especially since the Second World War.

Other than in the Arab world, dromedaries have always been rare in Central Asia, except in the southern areas. Instead, Central Asia without horses is unthinkable. Horses are the only means of private transport in this vast region where motorization has always been minimal, notwithstanding huge efforts to improve the infrastructure in Soviet times. For Central Asian livestock-breeders the horse is an essential means of communication, in order to keep the herds assembled, as well as for maintaining relations with neighbours and family. The horses are well adapted to the climate, they are supposed to graze the whole year round without being given additional fodder. The worship of the horse reflects its crucial role in all Central Asian societies (Rijk, 1993/1994:69; Stadelbauer, 1973:54 and Jettmar). This can be compared to the role of the dromedary in the Arab world. Numerous proverbs and poems reflect the veneration for this animal and the status of the owner of (large) herds (Pellat, 1971:665) as compared to sheep.

Different as well are the secondary activities of the livestock-breeders. In the Arab world, in Iran and Afghanistan, fields used to be prepared and sown before summer grazing started and the harvest was collected after the return. In Central Asia, however, only those who had lost their herds or could not afford to buy horses would be active in agriculture. Families who could afford to hire workers would engage in agriculture as well, since harvest time and migration of the herds to the summer pastures occurred in the same months. All family members were then needed to herd the animals.

The climate determines the type and form of the dwellings of the herdsmen in the two areas under discussion. In the North African and Arab world the vaulted black tent made of woven goat's hair is preferred and used. In Central Asia it is the dome-shaped *yurta* which is better at resisting the harsh winds of the steppe. In summer the heavy felt 'walls' of the yurta can be easily removed and, if necessary, replaced by reed mats to facilitate the movement of fresh air, whilst in winter extra layers of felt can be used to withstand the cold. In the following chapters, these famous yurtas will be discussed in more detail.

Where sufficient wood was available, notably in mountainous areas, the Kyrghyz used wood to construct log cabins instead of erecting yurtas. In Kazakhstan caves dug in the earth were also used as dwellings.

Since the advance of Russia into Central Asia, and especially after the Revolution, the traditional way of housing of livestock-breeders changed. Together with collectivization, living in a yurta was discouraged and a standard type of Russian-inspired dwelling promoted, complete with stables. These Russian-inspired dwellings are made of bricks and stucco for the walls, plywood and dung for the ceiling and a mixture of cement and asbestos for the roof. In the more southern areas adobe and sun-dried bricks, wooden poles and straw, are used for the construction.

Agricultural policies and privatization

Within modern extensive animal husbandry practices, a tendency is discernable for privatization of the herds. The Ministry of Agriculture promotes the breeding of the original Kazakh sheep, the Yedelbay, through privatized livestock stations. One of them is a limited liability company, called 'Assyl', comprising of forty kolkhozes.[5] They concentrate on Yedelbay sheep, a kind of sheep that will have to replace the Merino type in due course. Yedelbay sheep are able to look for food the whole year round, also in subzero temperatures, which is impossible for the Merino sheep. Only when the snow level is too high do these sheep get additional fodder.

The ministry also promotes the breeding of Karakul sheep in the west of Kazakhstan, where the Merinos do not prosper due to the hot climate and the lack of water resources. Here the Tzigay type is bred as well. Tzigay sheep produce very soft wool, used for cloth.

The breeding of dairy cattle is relatively new. They have been introduced to produce milk and cheese.

Recently some, not many fully private enterprises have come into existence, headed by former party-officials. In 1994 in the area of Kegen, near the Chinese border in Kazakhstan a former party-official had turned completely private. He owned three livestock stations and large herds of sheep, though the exact number of sheep could not be established. Interlocutors said somewhat maliciously that he had turned into a real *bay*, the title of the rich people before the Revolution. He was not only active in animal husbandry, but also owned a shop where imported foreign goods were sold.

Since middlemen owning trucks are appearing in the market, to transport, purchase, and sell private herds, hides, and wool at better prices than those of the state organizations (kolkhozes), this tendency of privatization will be strengthened.[6] Wool will be exported to China and Europe, although European buyers complain about the faulty selection methods of Central Asian wool where general selection standards of fibre length are not maintained.

Privatization also takes place whithin kolkhozes, where herds partly belong to the collective enterprise, partly to private owners connected or close to the kolkhoz. Stadelbauer (1970:181) quotes data from a kolkhoz which, in 1961, owned 560 goats and sheep collectively and 10 470 animals in private property. In Uzbekistan about 200 kilometres west and southwest of Tashkent, 40% of the herds were private property (1993) and managed by herdsmen belonging to the kolkhoz. The animals belonging to the kolkhoz were sold to the state and those belonging to private persons were put on 'the market'. It was said that this 'market' in Tashkent was the main buyer. In Kazakhstan in the two kolkhozes that we visited, the same percentage was mentioned. All members of kolkhozes visited in 1993 and 1994 said that between 40 and 60% of the animals were privately owned.

Life in a modern kolkhoz

In the kolkhoz Lenina in the Kazakh town of Uzyn Agach, west of Almaty, private ownership must have existed for a long time. One of the kolkhoz workers, Malik Sugan, is responsible for 2 000 sheep on the kolkhoz. Together

In the outermost southeastern part of Kazakhstan in the Taldy Gurgan oblast, we visited a kolkhoz which was privatized in 1993 through the sale of vouchers to the workers. The kolkhoz was split up into an agricultural section and an animal husbandry section. One of the (150) workers of the agricultural section explained that since 1992 the objective of the kolkhoz, in fact the company, was to produce barley, sugar-beet and potatoes. The production was partly for their own use, partly for sale. Of the total production of the company 50% had to be delivered to the state.[7] All assets had been given to the former members of the kolkhoz, now co-owners of the company. In case of positive results, 25% of this profit must be paid to the state. For the purchase of materials 328% of yearly interest must be paid to the supplier of capital.[8] Our spokesman now owned four cows and 300 sheep. Barley straw and a small stable behind their own house provided food and shelter for the cows. The company owned a stable for all the sheep of the fellow workers of the company. Herdsmen were hired by the private owners, to guard the animals for a salary of 550 Tenge per month.[9]

We were struck by the form of the kolkhoz house, situated at the outskirts of Kegen, which was a duplicate of the houses of the veterinarian of a kolkhoz 400 kilometres to the west and to that of a worker at a kolkhoz northwest of Usht Obe. Just as the buildings (house and stables) on the farms in Kazakhstan and Uzbekistan are alike. The only difference is the furnishing. Depending on income and family relations the outfit of a house, often made in Russian factories, will contain: more or less modern furniture, the television set in the guest room where visitors are received, the wall-hangings imported from China and the bric-à-brac from eastern European countries. The furnishing of the houses or apartments in towns and villages does not differ much. People, family and guests, will sit on carpets on the floor during lunch or dinner as it is done in the yurta. In the towns tables and chairs of 'western European' height will be used.

At the countryside – in farms and in villages – a visitor might see an ancient, traditionally made knotted and/or felt carpet. But most carpets and blankets are manufactured in factories and cannot even match the quality of the traditionally handmade carpets. In the yurta the interior will not show much of the ancient treasures of embroidery and weaving. Modern furnishings and bedding are the rule.

In Ushtobe, south of Lake Balkhash, the only yurta factory of the former Soviet Union is situated. The yearly production, 16,000 yurtas, are easily sold to clients in Kazakhstan and abroad.

Kazakh children in Uzyn Agach playing knucklebones.

with his two sons he runs the farm, which is situated a three hours' drive north of the town. In winter he stays with the animals near or at the farm, which is one of the subdivisions of the kolkhoz. Once every two days, and if necessary once a day, a truck will bring water for the family and, of course, for the sheep. Also bricks, cement and timber for the repair of the livestock station, stables and the house, fodder, provisions like sugar, flour for the bread which is baked once a week, and coal for heating, are transported over the partly paved, partly dust road. Meat is no problem: sheep and horses abound. And horse meat, more than the meat of sheep, is still highly valued in Kazakhstan and Kyrgyzstan, as it used to be. Whenever a guest arrives, a sheep will be slaughtered. When the same guest returns within a short period, horse will be served.

The members of Malik's family, working in agriculture and in animal husbandry, mutually assist each other with products. In the villages, as many agricultural products as possible are grown and preserved for use in winter and spring. Vitamin C is lacking in a country where the food supply fails. Butter, *koumiss* (fermented mare's milk), meat and wool, are exchanged for food containing vitamin C, for instance, for apples from Almaty, which literally means: father of the apples.

At the end of the spring, around 15 May, Malik will travel with his sheep for more than a week to the Alatau mountains into Kyrgyzstan with his family. He will then cover a distance of some 200 kilometres. The spot where he will install his yurta (in fact the yurta of the kolkhoz) and spend the summer till the end of September, traditionally belongs to his family and clan. His father and grandfather were herdsmen, too, and obviously rich, had many horses, which were more valued than sheep (as the Kazakh saying goes: *The horse is the plow of mankind*). When Malik was twenty years old, studying in Almaty to become a veterinarian, his father died. The family urged him to return, in order to replace his father and 'guard the family fortune', which is the herd. At present, half of the herd belongs to him and his sons and the other half to the kolkhoz.

During the journey to the summer pastures in the mountains, Malik and his family will pass the nights in the farms of his collegues. Those farms are situated at regular intervals between his allocated farm and the Alatau mountain range. Formerly a camel (which was only used for transporting the yurta), afterwards horses, and presently a kolkhoz truck will bring the yurta and the personal belongings and provisions to the traditional summer camping ground.

Apart from his income as a kolkhoz member, Malik earns money from the sale of animals on the 'market' and from the wild onions harvest in April and May. As a kolkhoz member he earns one Tenge per sheep per month, which is paid in kind, like flour, sugar, cloth, coal, etc. During shearing time and the lambing season, he earns more than the other months.

The kolkhoz of which Malik is a member, owns 45 000 sheep but manages 80 000 animals. The kolkhoz is run by a director, a veterinarian with twelve laboratory assistants, an economist, a book-keeper, a secretary and 160 herdsmen, half of them assistants. Sometimes, especially in spring, more hands are needed. This is also the case when the winter is very harsh and the animals are not capable of finding their own food. Then hay and straw must be distributed: three tons of fodder per thousand sheep.[10]

Depending on the type of work, other members of the agricultural section of the kolkhoz are called in. But wandering specialist herdsmen are also hired. Those herdsmen offer their services whereever they can be used, in Afghanistan, China, Uzbekistan, or elsewhere.

In addition to winter stables with storage facilities for fodder, machinery to feed the animals and a barn used for the shearing of the sheep, the kolkhoz is also equipped with a laboratory. The veterinarian looks after the horses and the sheep of the kolkhoz. A problem is, as he said, the availability of medicine which was very hard to get hold of, since the regional supply office is badly stocked. A special section of the kolkhoz is active in harvesting grass and shrubs in the mountains and on the steppes to be used as fodder in winter. In former times a long winter could mean the death of a large part of the herd due to the lack of fodder. The organization of fodder supply has changed considerably since that time. After the winter the herds are prepared for their stay in the summer pastures. The sheep will be sheared, cleaned, medically checked and medicines administered.[11] Cripple animals and a number of rams will be sold to the meat industry in Almaty. In autumn the animals are counted. As a rule, fifty out of every hundred kolkhoz sheep will be handed over to the state for the processing of meat. The quality of the pastures, however, will determine the exact percentage.

In former days, around the turn of the century, the grandfather of Malik would always use pastures closer to the mountains. But after the arrival of the colonists from the west, the marshy area around Uzyn Agach was drained and agriculture replaced animal husbandry. The colonists were Russians, Byelorussians, Balts, Ukranians, and later, among others, (deported) Koreans and Wolga-Germans. On being asked, Malik Sugan commented that in the kolkhoz nobody had complained at that time about the withdrawal of pastures for agricultural use. However, he feared problems in future, because of further privatization of pastures and arable land. And he realized that the times when the horses were the most important asset of the herdsmen, would never come back.

Conclusion

Our short stay in Central Asia and the discussions with the people we met, cannot lead to firm conclusions, only to impressions. Impressions of an area in turmoil where the young see possibilities at last. Five years ago the intelligentsia were somber. *'Gorbachev is just another Brezhnev'* as one of the interlocutors said in Dushanbe. Now, however, new challenges must be faced.

Central Asia knows serious environmental problems. These join the three issues that livestock owners worry about continuously: the availability of water, of fodder and of veterinarian services for their animals. The size of a breeder's herd depends on these factors. Sheep need water every 24 hours. Through the digging of wells the range of the herd movements could be extended, and therefore more fields used as grazing areas. Kolkhozes are and were responsible for the distribution of grazing facilities for the herds. The continuous urge by state officials to increase agricultural output has diminished possibilities to increase the size of the herds. No information on the present state of this potential dispute is available.

The present lack of medicine is detrimental to the health of the herds. Stomach and intestinal illnesses are the main factors which influence the size of the herds negatively.

The environmental hazards in Central Asia are presently great. Dust storms spread polluted materials from the shrunken Aral Sea over agricultural and pasture lands. In the pasture lands that we visited in both Uzbekistan and Kazakhstan, many patches of salt were discerned. The effects will be known in the future for both the inhabitants and the animals.

Times change and so do the opinions of young Kazakhs and Russians alike. Although the future of the first are brighter as the government has taken, in 1993 and 1994, many measures which threaten the existence of the minorities (prohibition to buy houses or apartments, obligation to learn the local language to qualify for jobs, in an area where Russian is the lingua franca). But these are the opinions of town dwellers.

In the countryside times are changing as well. Compared to the cities, life is harder, less money is available, food and other necessities are (much) more expensive, the quality of the medical services is receding. Livestock-breeders used to be confronted with party and state officials, with planners and economists, many of them Russians. Gradually they are and will be confronted by market conditions, market conditions in the Western sense of the word. Competition by Central Asia with the perfectly organized South African and Australian producers will be hard. They know that they lag behind. They also know that the environmental and economic conditions are worsening. Striking, however, is the positive outlook they have, based on their profound knowledge of the country and its potential for animal husbandry.

Notes

1. Mani, born in 216 AD, was the founder of Manichaeism, a religious system in which gnosis is dominant and Christian elements can be discerned. This faith spread from the Near East to North Africa in the west, and to east Turkestan and China in the east. It eventually disappeared in the 14th century under the growing influence of Islam.
2. Nestorian Christianity, founded by Nestorius, at the end of the 4th century AD stressed the dualist nature of Christ. It spread from Syria far into Central Asia. Nestorians were persecuted by the Mongols after the latter's conversion to Islam.
3. Bennigsen (1985:78-79) gives details of the different tribes, the geographical boundaries of the tribal areas and the tribal federations.
4. In kolkhozes, cooperative units independent of the state, each worker receives his salary once a year, with an additional small amount every month, often partly in kind. This amount can be different per person. In sovkhozes, which are state organizations planned and budgeted by the Ministry of Agriculture, every worker receives the same amount, which is paid each month, with an extra bonus at the end of the year, when the income of the sovkhoz permits. A sovkhoz is managed by a director, assisted by administrative personnel. Sections of the sovkhoz like those for animal husbandry, agriculture, services, maintenance of materials are managed by leaders. The animal husbandry section has, on its payroll, among others one veterinarian per 6 000 sheep and some zoological workers. The leader of this section will manage the brigades (40 to 50 persons). The sub-brigades number 5 to 20 persons. In theory a herd consists of 500 to 700 sheep and is managed by three herdsmen.

5. 49% of the shares belong to the workers, 51% to the state.
6. Middlemen owning a truck normally take 50% of the value of wool, the owner of the herd gets the other half.
7. No price indication could be obtained. In Uzbekistan members of a kolkhoz complained about the very low price paid by the state for sheep delivered to the slaughterhouse. In the 'market' the price was much better and in fact generally more than double the price paid by state organizations. Here, too, the obligation to supply existed.
8. The average inflation is 25% per month. In the *Deutsche Allgemeine, Zeitung der Russlanddeutschen* of 14 May 1994, a complaint was expressed about the time needed by banks to remit money in Kazakhstan which on an average took twenty days.
9. Since the end of 1993 the ruble has been replaced in Kazakhstan by the tenge. The value of the tenge as compared to one US dollar was, in mid-April 1994, 31 Tenge and at the end of May 1994, 48 Tenge. The salary of the herdsmen and other kolkhoz workers is in general between 200 and 250 US dollar yearly.
10. The traditional Kazakh sheep, the Yedelbay, is nearly everywhere replaced by the Merino type. The Yedelbay is capable, even under thick layers of snow, of looking for its food. Near Almaty a privatized kolkhoz managed by former civil servants of the Ministry of Agriculture has a breeding programme for this type of sheep.
11. According to a veterinarian the most common sicknesses are brucellosis and tuberculosis (for more details see Gatenby, 1991:81 et seq.). For the time being sufficient medicines are available. They are delivered by the relevant state organization. Interlocutors said that 3 to 4% of the herds die because of sickness and climatic conditions yearly.

2 Nomadic year cycles and cultural life of Central Asian livestock-breeders before the 20th century

Tatjana Emeljanenko

Historical peculiarities of the economic activities of the people of Central Asia, as well as the geographical setting, can be distinguished in the people's way of life and surrounding material culture: their settlements and houses, means of transport, food, etc. From ancient times Central Asia knew a great variety of such regional economic and cultural peculiarities. These resulted from the natural conditions, with its combination of vast sand and clay deserts, the steppes – well fit for animal husbandry – and the mighty mountain ranges where, in the hills lying at the feet of the mountains, and in the river valleys and deltas, ancient agricultural centres appeared. This chapter focuses on the way these peculiarities had developed during the 19th century in Central Asia.

Anthropologists distinguish in this period three main cultural and economic characteristics of Central Asian society:

1. The sedentary oasis population leading an intensive agricultural economy;
2. The semisedentary or seminomadic population, occupied in both animal husbandry and agriculture;
3. The nomadic livestock-breeders.

This subdivision, however, is general and pretty conventional. Indeed, each people (tribe) – the Uzbeks, the Tajiks, the Turkmen, the Kyrghyz, the Kazakhs, the Karakalpaks – could be characterized by one of those three socioeconomic features. However, specific ethnic and local groups within one region could belong to different cultural-economic societies as well. Besides, agricultural and animal husbandry traditions were an integral part of the culture of all peoples of Central Asia. The subdivision also underestimates the fact that the settled agriculture peoples and the steppe nomads could not exist without each other. There have always been very close political, economic and cultural contacts. These centuries-old relations played a very important role in the creation of the material and spiritual cultures of *all* peoples of Central Asia, although these cultures were ethnically specified.

In this article we will describe ancient nomadic animal husbandry cultures. We would like to show not only the role which animal husbandry played in the life of a nation, a tribe or a single person, but also how it defined their way of life and world-outlook, with a focus both on common characteristics of animal husbandry, as well as peculiarities of local groups.

The nomadic year cycle of the Kazakhs and Kyrghyz

As stated in the introduction, animal husbandry was widespread in the steppes, mountains and deserts that were inhabited by the Tajiks, the Turkmen and the Kyrghyz. However, only the Kazakhs, who lived in the western and southern steppes of Kazakhstan and the Seven River (Semirech'e) region[1], and the Kyrghyz who occupied the mountain plateaus and valleys of the Tien Shan, kept nomadic traditions up to the end of the 19th century. Over the centuries, a rational way of nomadism – changing pastures, ways and methods of herding – had been worked out and Kazakh nomads travelled long distances with their herds, mostly along the meridian. But even they used to have plots of arable land, somewhere in the environment of their dwellings. In other regions, however, this nomadic tradition was stronger, combined with agriculture, and livestock-breeders led a seminomadic way of life.

In summer the Kazakhs moved far north and in winter they went south. The radius of those trips ranged from about 200 to 1 200 kilometres and depended on place and size of the flock. All herding territory consisted of four types of seasonal pastures: winter, spring, summer and fall pastures. The winter pastures were located on places protected from cold winds and snowfalls by mountains, thickets or sand hills. The summer pastures used to be located in feather grass steppes and lake and river valleys. In the mountain areas of southern and eastern Kazakhstan the so-called 'vertical' system of nomadic life (*transhumance*) existed. In this system the summer pastures were located on higher mountain meadows, whereas the winter pastures were located at the foot of the mountain, protected from cold winds and blizzards.

Also, the Kyrghyz of the Tien Shan practised this transhumance. Their nomadic trips varied per region: from only some tens to 100–120 km, sometimes even 150–200 km. A lot of summer pastures were located high up in the mountains, up to 3 500 meters, close to eternal snow and glaciers. In fall, when the cold began, the breeders moved down to the fall pastures. These had often been used in spring as well and the grass had time to recover.

Each type of herds needed specific pastures, with special relief and also a special kind of grass. Up to the middle of the 19th century the Kyrghyz preferred to breed horses. Horses were not only useful for economic reasons, but played a role as well in the numerous civil clashes and military raids. For this purpose the so-called 'Argamak' horses were highly rated.

Local horses were pretty small, with much staying-power and strength. The nomads treated their horses in a very good way, for nomadic life and animal husbandry was impossible without them. This is reflected in folklore. In Kyrghyz legends the horse was shown as a protector of sheep, and the same idea existed for the Kazakhs. In one Kazakh epos, the *Shubarat* for instance, the horse of the hero Alpamys very often helps his endangered master and provides him with daring advice (Kazakhskiy, 1963:21, 222, 410).

Camels were treated in the same respectful way by the nomads, although they, like sheep, were less important. From ancient times yaks were reared in the mountains of Tien Shan and Pamir Alay. In the second half of the 19th century, the political situation in this area stabilized, and the breeding of sheep became very important. The number of animals increased enormously.

Each nomadic group had its own territory and routes, which passed from generation to generation. The nomadic communities consisted of small families that used to come from one and the same ancestor, and usually belonged to the same related group. Members of other clans rarely joined in, for this was considered to be bad, according to the local saying: *he, who separates from his relatives, will be eaten by wolves.*

The number of households in such communities could be different and depended on the number of animals. The nomads gathered to move from one place to another, for together it was easier to herd and protect the animals, to overcome difficulties en route and to use the pastures more effectively. Rich owners (so-called *bay*) could move by themselves; sometimes poor relatives might accompany them, but only if it was profitable for the 'bay'.

The conditions of winter herding, that started in November-December, were very bad. On territories close to the winter camps (*kyshtoh*) the Kyrghyz and the Kazakhs only herded camels, cows, young animals and such horses that were only good for long-distance trips. Other horses and sheep were herded to far away pastures with less snow. The winter herding system was the same for all breeders in one region. For instance, livestock and sheep that could not get food from under the snow were herded to the southern slopes of the mountains, whereas horses were herded to places covered with snow. If the whole plot was covered with snow, the horses, that shovelled the snow aside with their hoofs, were allowed to go first; the rest of the herds followed.

During snowy winters the animals suffered famine. The steppe grounds, especially where the Kazakhs lived, were ice-covered. Whenever this happened, the whole population of a mountain village went out to the steppes to help the animals. They tried to break the ice cover with shovels and axes, and remove the snow. Nevertheless, the combination of ice cover and hunger often led to high death rates among the herds (*jut*). Jut happened once every ten to twelve years, sometimes threatening large parts, or even the whole of Kazakhstan. The legend of the 'hard years' from this twelve-year animal cycle is connected with those *juti*. The worst years were considered the year of the mouse, the year of the hen and especially the year of the rabbit, when the greatest juts used to happen.

After a cold and harsh winter, moving to the spring pastures meant a real holiday for all nomads. They stayed at these pastures from the end of March-April till May. In spring the animals gave birth and it was necessary to prepare the young animals to move to the other pastures. Also, sheep were sheared in that period.

After this, the nomads moved to the summer pastures (Kazakh: *jaylau*; Kyrghyz: *jayloo*) where they stayed until the end of August or the beginning of September. This summer period was crucial and the arrival at the summer pastures was celebrated. The village (*aul*) that arrived first had to invite the neighbours who came later. Such a meeting of *auls* was accompanied by meals and drinking, whereas the various entertainments and games bore a ritual meaning as well: they were intended to help the herds staying healthy and prolific.

The main concern at the summer pastures was the processing of the products of animal husbandry (milk, wool) and to have the animals additionally fed for wintertime: the life of the whole family depended on well-fed sheep.

When the fall came, people moved to the autumn pastures, where they stayed for about 2 to 2.5 months. They sheared the sheep, repaired tools and yurtas, sewed warm clothes, mated the animals, slaughtered some and prepared meat for the winter. In November or at the beginning of December, they returned to the winter pastures.

For the Kazakhs and Kyrghyz, animal husbandry was their main concern. They learned all traditional skills concerning the treatment of the animals from their early childhood. According to a Kyrghyz saying, every child should learn how to ride a horse before he learned how to walk, for a livestock-breeder spent most of his life on horseback: on nomadic trips, while protecting and herding the flocks, etc. Women and girls were involved as well. They guarded the herds at night-time. The cycle of shepherd's songs to frighten the wolf (*bekbekey*) was created by women.

Below, we will see that many elements and forms of this nomadic year cycle and work were typical for other livestock-breeders as well. However, only for nomads were livestock the main source of income, and this they kept for centuries, up to the beginning of the 20th century.

Turkmen livestock breeding and agriculture

Turkmen livestock breeding has always been connected with agriculture. Only a few Turkmen tribes – the Balkhan Ata, the Mangyshlak Chovdor and the Abdal – were just nomadic livestock-breeders. The members of most other Turkmen tribes were, according to their line of work, split up into breeders (*charva*) and settled farmers (*chomur*). The farmers lived in oases. They out-numbered by far the charva, busy with animal husbandry.

However, in spring part of the oasis population would move to the pastures to help their kin breeders. One of the authors describing the life of the Turkmen in 1880, wrote: *'one cannot call the Turkmen living near the Caspian Sea fully settled (...). Even charva are no real nomads. Their family and belongings stay in their villages (auls); only at the end of spring their wives and female workers move to the sands, to gather milk and to prepare products of it. The owners and the male workers shear the wool. At harvest-time the charva return to the auls and the animals feed themselves with straw left'* (Obzor, 1897:25-26).

Animal husbandry traditionally concerned the Turkmen tribes, and this considerably influenced their world-outlook and culture. For instance, the Turkmen-Saryk, engaged in agriculture for a long time, answered the question 'Where do you live?', by: *'khatab galasyndah'; we live in the fortress of the camel saddle* (Ovezberdyev, 1962:113). Sheep and camels (80% of all livestock), and horses were the most important animals in the Turkmen economy and household; cows were somewhat less important. Camels were used for transport, preferably the 'Arvan' dromedary, famous for its strength and staying-power. The main part of the livestock was being kept at grass, watched all year round by shepherds.

The character of nomadic life in different areas depended on the natural conditions and the location of the pastures. Routes and pastures were traditionally divided between the tribes and tribal groups (*tire*), and were determined by the location of wells and other springs.

The livestock-breeders of northwest Turkmenistan (Orazov, 1962:292-294) stayed for about half a year (from June till November) at the summer pastures (*yaylag*), which were located around the water sources. When the temperature changed, the herds were driven to the winter pastures (*gyshtag*), in most cases located in the valleys or in the sands, where it was warmer, with sufficient forage and fuel. In spring (March to May) all animals were driven to the plain spring pastures (*yaylag*) with succulent green vegetation. When the grass started drying away in the hot sun, the nomads moved to the summer pastures again. Every year this same cycle was repeated.

In Kara Kum (König, 1962:246-248) people lived a semisedentary life. The herds were grazing on fairly small territories. Lack of appropriate water sources did not allow the breeders to travel a long distance; besides, the desert pastures could serve the animals the whole year round, since every season had its own kind of vegetation. There was always something to eat. The Tekyns of the eastern Akhal areas (where the pastures were better and more water was available) were even leading an almost settled life around their wells. Only now and then, in the spring, they moved to the natural water reservoirs, full of melting water from the snow. The Tekyns of the western Akhal areas, however, could not completely settle, but had to move on three times a year, within a range of 20 to 25 kilometres from their settlements.

Turkmen traditionally considered the pastures as places of natural wealth, belonging to everybody. These traditions were similar to the customs of the settled farmers, according to which uncultivated land could not privately be owned. Pastures, however, only could be used if they had wells. Most wells were owned by a family or by a group of families, who, as a result, extended their user rights to the nearest pastures.

The smallest autonomous unit of Turkmen livestock-breeders consisted of such a group of related families. Each community – some ten to fifteen households, with 40 to 60, or 100 to 200 people – had a name or was called after the nickname of its main ancestor. They worked together in animal husbandry, and participated in digging and cleaning wells, paying fines, blood vengeance etc. In winter the communities disintegrated into small groups, consisting of some two to five households, whereas in summer they gathered in the neighbourhood of the traditional water sources, forming the so-called settlements (*oba*). These oba could consist of some 40 *kibitki* (nomad tents). Large oba were often subdivided into smaller ones, located some 200 to 250 meters from each other. Problems and disputes between the members of the oba were not solved according to Islamic law, but following the common law (*adat*), supervised by the council of family heads (*aksakal*) (König, 1962:254). Water, and the wells, were an important uniting factor.

Water sources, especially those man-made, played a crucial role in the relations between the Turkmen livestock-breeders, which had been developed in desert conditions. The system of livestock watering, the digging and locating of the wells, had been worked out over many centuries; related traditions passed on from generation to generation.

Wells were built by special craftsmen, experts in locating the right spot through marks of the condition of the ground and the vegetation. Since wells had to be dug 20 to 40 metres deep, or even more, the expert had to know his trade perfectly well, in order 'to feel' the water. This was mostly done by Iranians or Kurds, (rarely by Turkmen), assisted by Turkmen involved in that area. Deep wells were covered inside with stones or haloxylon bushes.

Most Turkmen wells provided water that was slightly or bitterly salty, fit only for animals. The summer pastures (*yaylag*) and villages (*auls*) mostly had fresh water, although in western Turkmenistan settlements existed, without any unsalted well at all. People were obliged to get water from elsewhere (Orazov, 1962:292-293).

The owner of the well (the person who had the well constructed) was highly respected by his neighbours. The well often bore his name. The community wells, built by several families, was named after the leader of the work. However, although all wells had their owners (one or several), according to custom no one could be deprived of the right to use a well. It was believed that the more water was taken from a well, the better it would become. Letting 'strangers' use the water was seen as a deed to please God. Yet, there were some rules concerning the taking of turns at watering: the owner of the well started watering his herds, next came his relatives, followed by others. However, if someone arrived at the well before the owner, he was allowed to use it for his animals, and the owner had to wait.

Nobody had to pay for using the well, but everyone who used it had to take part in its cleaning, and, furthermore, to help the owner with shearing and milking his sheep, repairing his house, etc.

The custom of mutual aid, born under the harsh conditions of nomadic life, was firmly rooted in every livestock-breeder's mind. Everyone knew that not observing these rules could have tragical effects, for the person involved as well as for the members of his (her) family and home.

The sheep of the livestock-breeders of one oba were gathered in flocks – 400 to 500 sheep each (sometimes up to 1 000 sheep) – which were herded by the shepherd (*chopan*) and his assistant (*choluk*). With the help of specially trained dogs the shepherd had to watch the herds, to prevent them from straying, becoming lost or dispersed. He also had to protect the herds from wolves and to render veterinarian help if needed (Vasil'eva, 1954:115). In the eastern part of the desert, where people had to get water from very deep wells, the chopan and choluk were assisted by a *suvchi*, who had to pull up the water. Mostly, the shepherd was paid in young lambs. The Turkmen-Nokhurly, however, provided him with milk. When in spring the young lambs were born, the shepherd, during fifteen to eighteen days, was allowed to take all milk left after feeding. In summer he could take all milk once a week; by the end of the herding, in fall, once in three days.

The shepherds were chosen from the community or were attracted from elsewhere. Usually, young men from relatively poor families, who were about to get married but did not have enough means to pay the bride-money, became shepherds. Others, however, were middle-aged or even old.

As the Turkmen economy was more complex than the Kazakh or Kyrghyz nomadic society, this role of the hired shepherd was more elaborated as well. Nevertheless, most Turkmen were closer to Kazakh or Kyrghyz nomadic livestock-breeders, in their way of animal husbandry, than to the seminomadic or partly settled breeders. They did not store the forage. In their herds, camels and lambs, which were able to travel on very long distances, prevailed. Finally, although most Turkmen were active in agriculture as well as in animal husbandry, these activities were kept separate.

Animal husbandry and permanent settlements: Karakalpak, Uzbek and Tajik

In seminomadic societies, agriculture is predominant, but closely related to animal husbandry and with related activities like hay procurement, and keeping draught oxen. Around the turn of the 20th century, various peoples in Central Asia lived a seminomadic life. This applied to the inhabitants of the Amu Darya and the Syr Darya deltas – Karakalpaks, Turkmen and Uzbeks from northern Khorezm of the oasis of Bukhara – as well as to Turkmen, Uzbeks and Kazakhs who lived in the Fergana valley. In the mountain regions of Central Asia, the people of the seminomadic type were: the Kyrghyz in the area around Isyk Köl Lake, the Tajiks and the Uzbeks from the mountains of Baysun and Kygystan, and people in the Pamir.

Most Karakalpaks, the Uzbek-Aral, a part of the Turkmen and the Prisyr-Darya Kazakhs were descendants of the tribes and peoples that already had a complex economy in the early and late Middle Ages. These inhabitants of the vast delta regions, the lake and river districts, probably inherited their archaic traditional complex agriculture-breeding-fishing economy from the same tribes which inhabited these areas in the Bronze era.

Karakalpak The Karakalpaks were 'forced' to lead a nomadic life, because of the livestock and because of the natural conditions of their environment: at the lower reaches of the Amu Darya and the Syr Darya, the channels often changed and flooded new areas, so the people had to leave them and go to drier land (Etnografiya, 1980:17-22). According to a physician, exploring the northern regions of the Amu Darya river, during the floods *'the Karakalpaks stoical accepts the well-known caprices of the river; he builds a hummock in the water or a raft of reed and brings his tent over there. After that he drives his herd to places that are still dry and patiently goes on living and patiently waits in the middle of immense reed-lands and millions of mosquitoes till the water falls.'* (Avdakushin, 1892:15). Notwithstanding the heavy natural conditions, yet the Karakalpak way of life and culture were determined by animal husbandry.

Oxen and cows were the most important; this is affirmed by the fact that the dowry (*kalym*) could only be paid in oxen and cows. In other livestock-breeding communities the kalym consisted of whatever livestock available. The number of animals was determined by the family income. Horses and sheep, however, were less important. Oxen were used to plough the land, to turn the watering wheel (*chigyr*), in threshing (they trampled the ears of corn and beat the kernels out of them) and, finally, they were the main means of transport: they were harnessed to a cart (*arbah*) with which the Karakalpaks did all their nomadic trips.

In wintertime the Karakalpaks brought their livestock into enclosures made of branches and reeds or in closed accommodations located near the dwellings. Dug-outs for horses and cows were built. The forage was laid in for all livestock, with the exception of the sheep, that could feed themselves the whole year long, getting their food from under the snow in wintertime. Cows and oxen lived on dried reeds, straw, rice, wheat, millet hay and other sorts of hay. For camels some steppe plants were laid in. Horses were fed with wheat hay, barley, jugara (the local variant of millet) and lucerne.

Karakalpak winter settlements (*kyslav*), as with the other livestock-breeders discussed here, consisted of members of related groups. These kinship ties were also relevant when pasturing the animals during the seasons. In spring the herds were driven to the pastures. Their location depended on the wealth of the livestock owners. The rich could leave some of their relatives, or hired workers, in the aul to work in the fields; they themselves went with the herds and the horses, often very far from the aul, to the Kyzyl Kum wells or to lakes and other water sources with enough reeds. Less wealthy people herded their livestock not far from the winter aul. The women and children joined the herdsmen, whereas the remaining men stayed in the settlement to cultivate the land. The nearest pastures (*jazlau*), were used in the summer, when agriculture needed most hands.

Tradition ruled that all families, with their tents and other belongings, had to leave the settlement in summer, though the distance of their 'trip' was at times not more than 500 metres (Kaul'bars, 1881:545). The Karakalpaks travelled in special carts (*arbah*) yoked by oxen. When the herds had to be driven along the bank of a lake or river, the families travelled by boats. The herdsmen were hired collectively, and each family gave them food and a place to stay. Each type of livestock – sheep, cows, horses – knew special herdsmen.

Uzbek The Uzbek way of animal húsbandry was totally different at the end of the 19th century. This difference was caused by the natural conditions of their environment and old economic traditions (Karmysheva, 1969:44-49).

Only the Uzbek-Kungrat who lived in desertlike areas, especially the basin of the Sherabad Darya river middle stream, were real nomadic livestock-breeders. The livestock-breeders and their families constantly followed their herds to the pastures which were located within the borders of their kin tribal group. They used to change places every two to three weeks. The settled Uzbeks (and Tajiks) of the valleys and at the foot of the mountains were breeding their livestock on pastures close to their settlements (*kyshlak*) in all seasons. After growing wheat, horticulture, viticulture and various handicrafts, animal husbandry took the third or fourth position in their economy.

The Uzbeks, who lived a seminomadic life up to the 1920s, were the descendants of an ancient Turkish nation called the Mawerannakhrah and Uzbeks of Deshtikypchak origin. Sheep breeding was the most popular and widespread practice among them. According to tradition, they treated their sheep in a special way. They believed that sheep originally came down from heaven to earth. Sheep breeding was considered beneficial and shepherds were highly respected by everybody.

Besides sheep, horse breeding also took an important place. They were used for threshing and, like camels and donkeys, for the transportation of hay and corn. Cows and oxen were the main draught animals. Almost everyone kept goats. These were the animals typical for most seminomadic Uzbeks (Karmysheva, 1954:139-142).

The nomadic cycle of Uzbek livestock-breeders covered only small distances. In winter the animals stayed two to three kilometres from the settlements, which were located in river valleys, on the southern slopes and close to the wells and springs. Here they constructed sheds and enclosures. In early spring (before the lambing season), the Uzbeks left their villages and moved to the steppes or to the nearest hills. The large horse herds were driven to the southern desert areas, to enable the mares in foal – who had gone very skinny in winter – to recover and gain weight, whereas the ravines and steep slopes would be no obstacles for the newborn foals. The sheep, on the contrary, had to leave the desert pastures and come closer to the settlements, because when they started to give birth, the shepherds needed a lot of help. In summer when the steppes burned down, the flocks were driven to the summer pastures and by fall all returned to the settlements.

Each related group had its own pastures, although this was not officially legalized; ownership was acknowledged by clan traditions. Clan communities consisted of three to fifteen families or even more. The number of sheep in their flocks was 500 to 600 (in the mountain areas) or 600 to 1 000 (in the steppe areas). Each flock had two or three shepherds, who, in turn, received a couple of lambs and a set of clothes: a shirt, trousers, a turban, a wadded or woollen robe, a pair of shoes and two hats – a big one (*chalmah*) and a small fur one. They were also provided with food.

Only families with a very small number of animals stayed in the settlements in summer. Sometimes others returned for a while, (mostly men), to plant the rice, to take care of the watermelons, and to harvest.

Tajik The predominance of animal husbandry or agriculture among the mountainous Tajiks – who inhabited the Pamir and Gorno Badakhshan and the region west of it (the Kulyab, the Gissar, the Karotegin, the Darvaz) – depended on the relief of the land. Animal husbandry predominated in the higher settlements, close to the alpine pastures, with only some arable land. Traditionally, 'vertical' livestock driving, transhumance, existed. The way of herding, however, depended on the height of the settlements or the type of animals. Sheep and goats could easily live in this mountain landscape, but cows and oxen were important as well. Horses were not very popular and the number of camels was very limited.

The transhumance was necessary because the herds could not feed themselves all the year long. Therefore, they were driven from the alpine summer pastures to the winter pastures in steppes or valleys. The spring and fall pastures used to be at the foot of the mountains. In wintertime, during a large part of the year, the animals stayed in the enclosures near the settlements for six to eight months. Forage had been stored on beforehand, and it was provided very carefully: the animals were fed in the houses, to prevent them from trampling the food.

The population of a settlement usually only left it to go to the summer pastures. The men went first to repair or build the new enclosures, followed by the women, who worked the whole season in order to make dairy products. After the arrival of the women and children, the men returned to the settlement in order to cultivate the land.

The summer cycle and other traditions

Although animal husbandry varied in importance and practice, due to geographical and historical conditions, communities that lived far away from each other and with different ethnic and historic traditions, still showed a lot of similarities. The same goes for their systems of magical images and actions, connected with animal husbandry, and aimed at protecting the livestock, to make them healthy and productive[2] . To this we will now turn.

All livestock-breeders celebrated the beginning of summer or spring as the start of the summer cycle. Departure to the summer pastures used to take place on a Wednesday or Friday, which were considered to be good days. All the people washed themselves and put on their best clothes. The back, neck and head of pack-camels and pack-horses were covered with special holiday blankets, whereas the luggage packed on the animals was covered with carpets and colourful blankets as well. The same accounted for the carts drawn by camels and horses, as the Karakalpaks did. Pieces of the nomad tents (*yurtas*) and other utensils, were covered in special felt ornamental sacks (sometimes made of carpets). Among the Kyrghyz or Kazakhs the ceremony of the caravan procession was sometimes headed by young girls. They would ride on horse-back, dressed up in colourful clothes. Women, with their children, followed, seated on camels. After them the herds were driven. In other cases an elderly respected man rode in front. He was followed by the old men, who would lead camels with bells on their necks. Among some groups of Kyrghyz the camels were led by old women or young men. Among the Karakalpaks, Uzbeks and Tajiks, the herdsmen headed the train, followed by the caravan of people and the other animals.

Common among livestock-breeders as well, was the tradition to 'clean' the animals with fire. Before the livestock went to the pastures for the first time, it was driven between two big fires, in which salt or sulphur was thrown to increase the 'cleaning' power. With loud noises – guns shots (the Turkmen), or banging on metal objects – evil spirits were chased away. In the mountain settlements of the Tajiks, women fumigated the herds which came out of the enclosures with the smoke of burning grasses (*isirik, isvon* - Latin: *paganum harmala*). Before bringing the animals to the enclosures, the Uzbeks touched these with burning torches. The pastures were 'cleaned' with fire (torches, fires, smoke), as well.

At the start of the drive and after arrival at the summer pastures, a lavish entertainment was organized, in order to make the summer's herding a lucky one. Animals were slaughtered and dainties baked. The Uzbeks very often invited the mullah to such celebrations. He read the *Risolah* – the professional statutes, devoted to Chupon-Otah, the Patron of the sheep-breeders. During the passage, the caravans used to stop at holy trees and stones. The Mazar stopped at places connected with the names of Islamic saints and offered sacrifices – objects, or food; sometimes sheep were slaughtered. Thus, Islam and the ancient belief in magic, were united. Synchretism was typical for the spiritual culture of all peoples of that region, especially for the livestock-breeders.

Young herdsman and his family in their two-room farm, a stone house near Samsy, west of Almaty.

Kazakh herdsman

This yurta is characteristic of the steppes north of Almaty, Kazakhstan. From spring to autumn the yurta serves as a dwelling for the whole family. Since there is no well in the direct neighbourhood of the yurta, milk-cans are used as containers for drinking-water.

Soldiers of one of the counter-revolutionary White armies destroying an aul, a group of yurtas in Kazakhstan. This painting, one of a series of three with scenes from the civil war following the Revolution of 1917, had been painted in the thirties by an unknown artist. (Collection: Central State Museum of Kazakhstan, Almaty)

Nomadic livestock breeding is replaced by ranches of the type found elsewhere as well. In autumn the sheep are driven to their winter station in the valleys (Southern Kyrgyzstan).

Present-day livestock station near Uzyn Agach in the steppes of Kazakhstan. The farmers live in the small house and the yurta. In winter the sheep stay in the three large stables. On the left and the right of the house are enclosures used in spring and autumn. Coal from Karaganda and horse dung for heating, are piled up beside the yurta and the house.

The Registan in Samarkand with the madrasas of Ulugh Beg, Shir Dor and Tilya Kari. The beautiful Islamic ornaments are carefully restored.

Detail of a beautifully decorated woven strip, used to fasten the wooden frame of a yurta. Dromedaries, horses and shepherds are depicted in the characteristic colours of Central Asian tapestry (Yomud-Turkmen; end of the 19th century; height 44 cm). (Collection: Central State Museum of Kazakhstan, Almaty)

The Central Asian puppet theatre flourished until the beginning of this century. Two types of puppets co-existed: hand puppets or marionettes. Hand puppets resembled the Russian Petrushka theatre. The characters in the marionettes plays were influenced by the Turkish Karagöz theatre. In the Central Asian version real historical figures and symbolic personages played a role. The protagonist was Yuldam Yasavul, wearing a cap and a Russian uniform. He was involved in many adventures and found himself in awkward situations. In both villages and towns the theatres attracted many local and regional visitors.
The two Uzbek hand puppets represent Chinese soldiers. (Early 20th century - before 1908. Height 52 cm.) (Collection: Russian Ethnographic Museum, St. Petersburg)

Part of the dress of a Turkmen girl. The silver and gold-leaf amulets represent ancestors and forebears. The red colour protects against the evil eye (Turkmenistan, 19th century). (Collection: Russian Ethnographic Museum, St. Petersburg)

Many instruments and tools of the livestock-breeders were thought to possess magic qualities. The shepherd's crook, for instance, was highly respected. If someone by accident stepped on it, he (she) had to pick it up from the ground and kiss it three times; if someone stepped over it, he (she) had again to step over it, backwards. People believed that the person who stepped over the shepherd's crook would dry out like that crook, whereas a pregnant woman would give birth untimely. Even oaths were sworn on that crook (Karmysheva, 1954:114; Shaniyazov, 1964:77). Strings to tie young lambs, foals, camels, calfs and cows, or the reins for camels and bridles were equally believed to possess magic power. People neither liked to step over these objects and if someone did, this could bring him serious poverty (Khozyaystvo, 1980:132).

Many legends existed in relation to the magic qualities of the bones of certain animals. The Kazakhs hung the elbow bone of a lamb under the roof of their winter houses, or around a camel's neck, as a protection against wolves and thiefs. All the peoples of the region would tell fortunes by a lamb's shoulder blade: the cracks that appeared on it during burning, foretold the future. Various methods existed to cure sick animals, based on practice, on knowledge of the animals' peculiarities, as well as the powers of herbs and medicinal minerals. One often had to resort to magic actions, since all diseases were supposed to be caused by evil spirits. Thus, the Kyrghyz and Kazakhs went with the livestock to their cemeteries or their holy places (trees) in cases of livestock plague. They drove the animals three times around that place and stayed there for the night, praying to the ancestral spirits to cure them (Bayalieva, 1972:49). If the animals were ill, tradition ruled (among the Uzbeks, the Tajiks and the Karakalpaks) that witches were invited. They sometimes did a good job, as well as the mullah, who prayed and gave the owner amulets containing religious texts. Sacrificial rites connected with certain disasters were very popular as well.

Most rites and ceremonies connected with animal husbandry, kept up to the 20th century, were very ancient indeed. Some, although slightly altered, are still alive. They show the great role that animal husbandry played in the world-outlook, way of life and spiritual culture of the peoples of Central Asia.

Social bonds and communities

In the character of settlements and housing, of food processing, handicrafts, trade and other fields of material culture, animal husbandry left its traces as well. Of course, nomadic livestock-breeders had temporary settlements. The seminomads, with their permanent dwellings, used temporary settlements on the seasonal pastures. Such settlements were organized according to tradition.

As explained before, nomads travelled and lived in closely related groups. In the first half of the 19th century such communities even numbered up to hundred or more families (among Kyrghyz and Kazakhs), because of the always threatening intestine strives and raids. This joint herding by large communities led to a rapid deterioration of the pastures. Each new year, the large herds and flocks within a short time trampled down the grass, and people were obliged to drive them from place to place. However, at the end of the 19th century these communities diminished; at that time, a nomadic settlement used to consist of five to ten yurtas. Very rarely more than thirty households were involved.

Smaller villages used their pastures more effectively. They could stay longer in the same place. The size of the settlement and the number of households involved depended on the season as well. At the winter pastures, a small number of households joined together, in order to be able to use these pastures more effectively, whereas the summer pastures were often occupied by several settlements. In such cases, though, people kept a certain distance between each yurta. To this yurta we will now turn.

The yurta The yurta had a crucial meaning in the formation of the nomadic and seminomadic settlement. It was the traditional dwelling for nomadic breeders and for the peoples speaking Turkish and Mongolian languages. For the Kyrghyz and the Kazakhs, in the long-distant past, the yurta was the predominant type of house. Even when they settled in villages, at the end of the 19th century, it was often the only type of dwelling. In the Turkmen settlements there were both normal stable houses and yurtas, where young couples used to live and where the families preferred to spend the hot summertime. The Tekyns, Saryks, Goklens, for example, lived the whole year round in yurtas and adobe buildings were only built to keep food, or served as the enclosures for the cattle (Vasil'eva, 1954:135-136). Also, the yurta remained the main kind of house for the Karakalpaks, notwithstanding their agriculture and settled villages; and for some Uzbek groups (Karluk, Kypchak, Kungrat, Nayman, Mangyt, etc.) as well, in whose economy animal husbandry kept a prominent part.

Only the mountainous Tajik livestock-breeders never used yurtas. Even at the temporary summer pastures they built houses resembling their winter

Mounting a yurta is traditionally the work of women (Kyrgyzstan). (Photo collection, Central State Museum of Kazakhstan, Almaty)

Yurtas were traditionally transported on carts (Kyrgyzstan), camels (Kazakhstan) and horses. (Photo collection, Central State Museum of Kazakhstan, Almaty)

Presently, trucks are used for transporting the yurtas to the summer pasture (Kazakhstan).

settlements. More often though, they constructed simple cabins of stones or branches, called *kapa*. Such cabins were known to all the other peoples of the region. The Tajiks made it of branches and stones, built half under the ground. Sometimes parts of an old yurta were used to make these cabins: spokes in a dome-shape covered with felt or reed. The cabins served as a shelter for the shepherds. People with no possibility to get a new yurta also lived in a cabin. This happened, for instance, when the son of a Turkmen-Murgab family got married. As a start of his new life, he built a house of reeds (Vasil'eva, 1983:151-153).

Distinctions between an aul (the nomadic village) or a settled village were clear from the presence of yurtas. In an ancient aul, the yurtas were placed in a circle, with a sort of night-time enclosure for the animals in the middle (Potanov, 1949:57). The doors of the yurtas faced the centre. Even in the 19th century, among some groups of the southern and northern Kyrghyz, as well as in the steppe regions of Kazakhstan (the Priyrtysh steppes, the northern and the central areas), yurtas were placed in this position. Lambs were always kept in the middle, the calves and foals were tied up behind the yurtas. In the middle of the circle wooden stakes were sometimes placed, to rope the animals or hang up household utensils. In the smaller settlements the yurtas used to stand near each other, or in a semicircular form, the doors leeward. In other instances, the yurtas were placed without any special plan, following the relief of the terrain (Bartol'd, 1968:463; Shaniyazov, 1974:231).

Yurtas, widespread throughout Central Asia, used to have the same construction everywhere. There were only some small differences in the shape of the dome, the felt covers or other details, revealing ethnic, tribal, or local peculiarities.[3] Such variations could be essential, or only concerning a certain detail. We will give some examples of such differences.

A yurta's main framework consists of: a lattice that can be moved apart, the spokes and rim of the dome and the door frame. To cover the framework, mats of felt and reeds were used. In the past, the people or tribe could be distinguished by the decoration, shape and appearance of the yurta, the main distinction being between a yurta with a semicircular shaped dome, and a yurta with a truncated cone shape. This distinction depended on the way the dome spokes were bent. Yurtas with a semicircular shaped dome were typical for Karakalpaks, Uzbeks, southern Kyrghyz, most Kazakhs, and large ethnic groups of Turkmen – the Iomuds, the Tekyns, the Saryks, the Goklens etc. The second dome type was popular with the Kazakhs of the Semirech'e district, the Turkmen-Chovdor and the Ershahry, as well as in northern Kyrgyzstan

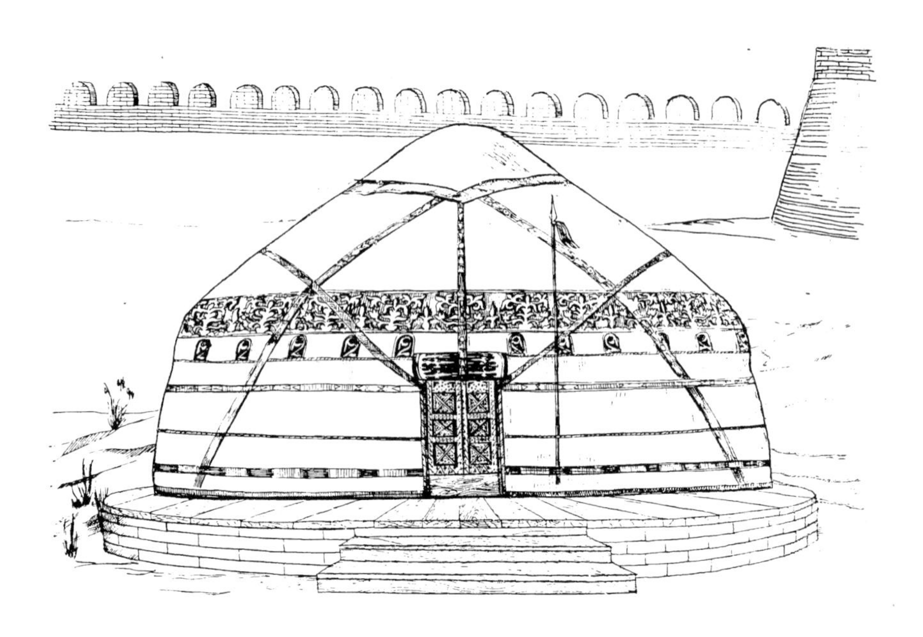

This large yurta from Kazakhstan belongs to a royal or feudal ruler. Its surface is 121 m^2; height 7 metres.

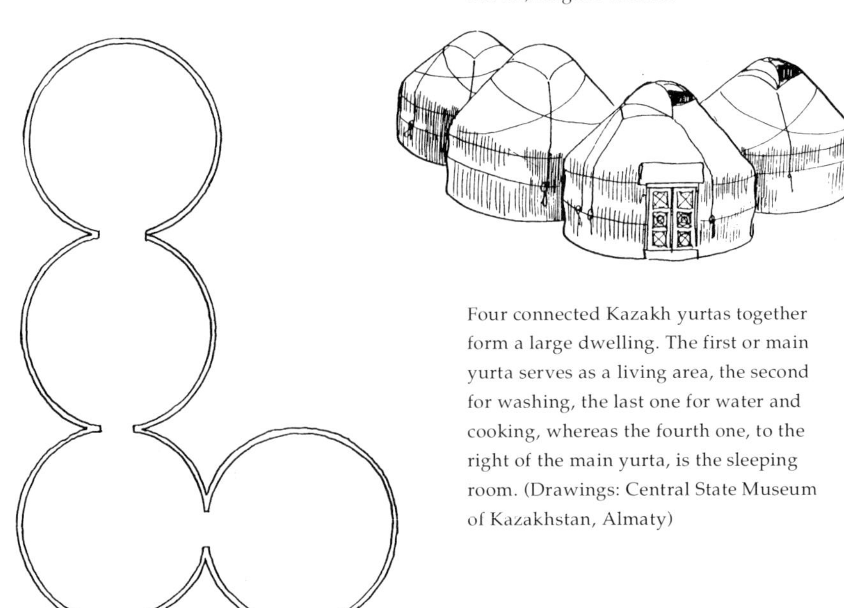

Four connected Kazakh yurtas together form a large dwelling. The first or main yurta serves as a living area, the second for washing, the last one for water and cooking, whereas the fourth one, to the right of the main yurta, is the sleeping room. (Drawings: Central State Museum of Kazakhstan, Almaty)

(with the exception of the Talas valley area). Domes as such could have a different shape as well, concerning the dome's rim, the height, the density of the lattice of the walls, the kind of felt, etc.

The Uzbeks, Kyrghyz and Kazakhs covered the wooden walls of their yurtas with a first layer of reed mats, covered, together with the top of the yurta, with felt. In the eyes of travellers at the end of the 19th century, the Karakalpak settlements in the lower delta of the Amu Darya and at the Aral shore seemed to be located in the 'kingdom of reeds'. They reported that all sorts of buildings were made of reeds over there, whereas the yurtas were only covered with mats (instead of felt covers). The Turkmen first covered the framework with felt and then covered these with reed mats. The Chovdors were an exception, putting mats in two layers. Most Kyrghyz covered the framework of the yurta with two layers of felt: the lower one covered the lattice; the upper one the dome. In the Talas valley and in some southern districts, the framework was covered from top to bottom with three to four strips of felt in one piece, which stopped 20-25 cm from the ground.

Outside decorations of the yurta were peculiar as well. Some groups of southern Kyrghyz and the seminomadic Uzbeks fastened the dome on the outside with large white ribbons. In most cases, however, ornamental bands were used: woven or embroidered, decorated with a pattern. The patterns were always connected with tribal or kin traditions, bearing a ritual meaning, related to the role the band (*bahu, booh*) played in the decoration of the yurta.

The shape and size of the yurta also determined its interior, with little variation within one region. Everthing in the yurta was placed along the walls in special wall bags, which hang on to the lattices or the dome poles. Although the yurta did not have any separate rooms, all the same, it was divided into several sections, each traditionally with its own meaning. The yurta knew a male and a female part, a place of honour and a place for the hearth. The female part was to the right of the entrance. Kitchen equipment, products and needlework were kept there. On the left of the entrance was the male part, where the corn and the horse harness were kept.

The hearth was placed in the centre of the yurta (with most Kazakhs, Kyrghyz and Turkmen) or closer to the entrance (the Turkmen-Ershahry, the Uzbek-Kypchak and the Karluks). The northern Uzbeks and the Karakalpaks of the Khorezm oasis had two hearths in their yurtas; one in the centre – for heating – and another on the right, closer to the entrance, to cook on. The place of honour was near the hearth, where the host received his honourable guests. The Kazakhs said: *When you enter the home, don't rush to the place of honour*, meaning that one should know his place. The best carpets and felts were in that place of honour. Most peoples also piled up their bedding (*juk*), felts and carpets at that place, where colourful bags with clothes and other things also hang. Only some Turkmen groups (for example the Ershahry), Uzbek-Kypchak and Karakalpaks kept their places of honour (*tohr*) empty. The stands with their belongings were located near the walls, closer to the woman's section. Many other differences in the interior of the yurta could be distinguished as well, related to the traditions of various ethnographical groups and differences in handicrafts, embroidery, felt and carpets; or the order in which the objects and utensils in the yurta were placed.

The role of the yurta in nomadic life was also reflected in the various rites and ceremonies that accompanied its placing. Constructing a yurta was accompanied by sacrifices, games and contests. Special attention was paid to the erection of the dome's rim (*shanrak, changarak, tundyk*). When this happened, the Uzbeks would slaughter a goat or a cock and pray to the holy protector of home and hearth. For the Kazakhs the dome was the family relic, and it was inherited from generation to generation. In serious cases of an oath, the Kazakhs' swore, looking at the dome (Vostrov and Zakharov, 1989:36). Karakalpak children were allowed to swing on a long rope, fastened to the rim that was attached to the dome spokes. Because of this swinging, the rim not only became more firmly attached to the spokes, the wall lattices and the spokes came closer to each other as well. Besides, it was a magical rite connected with a well-known fertility cult (boys and girls swinging in pairs).

Sometimes, also sacks with corn or stones are hung from the rim, because of their supposed ritual power. To be protected from the evil spirits and to make their home wealthy, the poles were smeared with butter and pepper, salt, onion, plants and grasses, and were hung up to the wall lattices (Etnografiya, 1980:39, 46-47).

The animals and the produce

Animal husbandry in Central Asia provided meat, milk, wool and hides as the main products. Only the men processed the hides, making footwear, harnesses, belts and other leather objects. Also from bones several objects were made by men (the Kazakhs were real masters of inlay of ivory in wooden objects: boxes, cases, beds etc.). The women produced wool and food. They spun threads out of wool, which were used for carpets. They also made felt out of woollen filaments. Centuries-old practical skills and a variety of magical rites and ceremonial actions played a role in the care of the livestock and the processing of the produce.

The livestock gave birth in early spring. Then everyone was busy, taking care of the young animals. Lambs' noses and mouths had to be cleaned of slime; when a lamb was first fed, the first stream of milk had to be wasted, before the lamb was allowed to drink. It was essential that the lamb took the udder during the very first hours of its life, for the first milk cleans the organisms. Sometimes, during the first days the newborn lambs were kept in the yurtas or in special earthen constructions. If there were Astrakhan lambs in the flock, the Uzbeks and the Turkmen selected the lambs of two, three days old: the majority of the male lambs was killed, to get the astrakhan, the females were destined to grow up and become the future productive forces. The male lambs of which the wool didn't turn white within ten days were slaughtered as well. Young lambs from other breeds were castrated (Abramzon, 1971:76-78; Shaniyazov, 1973:73-74).

During this same lambing season, the camels and horses were sheared, the young animals were marked with incisions on their ears (signs of ownership), the horses were trained and new herds put together. Right after shearing the horses and camels, the sheep were sheared. This started with a washing. The sheep were driven into a river or a special pond near the well. In small settlements, the people themselves sheared their sheep, in the larger ones neighbours and relatives were invited, or helpers hired.

This season demanded various rites. Almost everywhere the birth of the first camel or foal was welcomed with a ceremony; when the calves were born people had to take care that the cows did not eat the placenta, otherwise, according to the Tajiks, they could die, or get sick, as the Kazakhs believed (Andreev, 1958:119; Khozyaystvo, 1980:131). During the first days, the young animals were hidden from outsiders, to prevent them from being bewitched. Amulets were hung around the animals' necks.

The weeks when the animals gave birth used to be strictly planned. To prevent sheep from premature birthings, for example, the rams had to wear special aprons until the middle of August, or the beginning of September. The end of fall was the mating time. A 'lucky' day was chosen to start the mating. A mother with many children took the apron away. (The woman was chosen to promote the fertility and the birth of more female sheep). This happened at night so as to be protected from the evil eye, within the settlement (where the livestock appeared as well). Usually five rams were allowed in a flock of about 300 sheep, so that gradual mating was guaranteed. When all aprons had been taken off, the owner of the sheep took one ewe and poured some millet into her mouth, 'so that the increase was numerous' (Abramzon, 1971:79; Khozyaystvo:131).

By fall the sheep were sheared a second time. Autumn wool was more valuable than spring wool; the best felts, saddle clothes etc., were made out of the latter. The production of felt pieces started in the fall, but also in the summer a great deal of time was spent at making livestock breeding products. The felts took a considerable place in the economy of especially the nomadic Kyrghyz and Kazakhs. Felt was not only used for covers for the yurta, but in its interior as well: ornamented felts covered the floor, bags and sacks were made of felt, etc. (Antipina, 1962:22-40). The Kyrghyz and Kazakh women from different families traditionally worked together. The work was hard and demanded a lot of physical power. Collaboration in the production of felt was also common for other peoples in Central Asia, especially when large pieces of felt had to be made.

Milk and food In summertime the women also collaborated in food processing. This was considered a crucial activity of cattle breeding, and the women's respected position in society was closely related to their role in food processing from the cattle. They milked cows and worked up milk products. These products, that were processed in summer, provided the main food during the whole year. The food of the cattle-breeders was very closely related to the climate and the nomadic way of life. All food was divided into two categories, characterized by its energy components: 'hot' and 'cold'. The 'hot' products were, for example, mutton and mutton fat, horse fat and meat, and cereals; the 'cold' were: beef, horse fat, melted butter and oil, goat meat, most milk products and bread (Shaniyazov and Ismailov, 1981:120-121). According to this division, people made up their daily and seasonal food allowance. In general, meat was eaten less often, mostly as holiday food in wintertime or when important visitors arrived.

In the large settlements the women's cooperation in food processing was necessary, because for a single family, the work was too hard to cope with. In the small settlements mutual aid had another meaning. Women of the mountainous Tajiks and some groups of the Uzbeks (in the Fergana valley), with only a small number of cattle, in turn got all the milk gathered that day, during a certain period, to enable them in turn to make butter or to ferment it (Peshchereva, 1927:49).

The sheep and goats were milked two to three times a day, the rest of the livestock three to four times a day. The mares were milked for the first time about a month after the birth of the foal, five to twelve times a day. The camels had to be milked very often, too. Cows were milked as soon as the calf had gone to its mother; shortly after the first milk came, the calf was driven away.

All peoples of the region had the custom to make a ritual dish of the beestings (the first milk of cows, sheep and goats), two to three days after the birth. The Kazakhs, for example, mixed the beestings with water or milk, added salt, poured this into a sack made of ram's stomach or duodenum which was then immersed in a boiling meat broth. In thirty to forty minutes the beestings had curdled and became a white mass. It was cooled, cut in pieces and given to the guests together with meat. Family members were the first to eat and then the guests wishing them good luck: *let your home be wealthy, let you have a lot of milk and let the newborn animals and their mothers be healthy and strong* (Argynbaev, 1975:204). The mountain Tajiks boiled the beestings in a pot and only women and children were to eat it, dipping some special flat-round cakes made of grain in it (Andreev, 1958:119).

Central Asian peoples never drank fresh milk. Without exception, it was first boiled and then fermented; they made butter and various kinds of cheese out of the fermented milk. Each people used their own traditional methods. For the ferment they all used whey, pieces of cheese, melon seeds, branches of juniper, silver rings, coins etc. Where the horse breeding prevailed, especially with the Kyrghyz and the Kazakhs, the *koumiss* was very popular (koumiss is fermented mare's milk and is a real nutritious and curative product). At celebrations, this drink was always offered, especially at marriages and funerals. It was supposed to contain magic powers, so the joint drinking of koumiss was thought to be good for everyone.

A very similar drink was made out of camel's milk. One camel gave ten to fifteen litres of milk, and could be milked almost ten months a year. However, to camel milk other powers were attributed. The Karakalpaks and the Uzbeks of Khorezm, for example, thought that one got grey hair from camel's milk. Also camel meat got a special treatment. It was not given to pregnant women, for fear that childbirth would be delayed. Nevertheless all peoples also linked the camel in a positive way with fertility, and even nowadays women try to pass by a camel, or to get under it, when a childbirth is delayed (Snesarev, 1969:318).

Butter from sour milk was made in a special churn, of which various

A Kazakh woman milking a mare. Mare's milk is used to produce *koumiss*, a favourite drink in Central Asia. It is slightly alcoholic and tastes like sparkling buttermilk. (Photo: Central State Museum of Kazakhstan, Almaty)

The interior of the yurta 'reading corner' on a summer pasture. In the Central Asian republics many Soviet programmes were directed at eliminating illiteracy (Kazakhstan, Semi-Palatinsk Region). (Photo: Jamburg B.7 and Konoaloz A.B.)

types existed. The seminomadic and the semisettled people preferred a wooden or ceramic one; the nomads used a leather one. One way of making butter among the nomadic Kazakhs was very interesting: they churned it while riding 'in full tilt', a leather bag fastened to the saddle. The most popular cheese was *kurt*, made of dry curds. It was full of calories and was basic in many dishes (sometimes it was just mixed with water and pieces of bread). Kurt could be kept for a very long time and was stored (like butter).

Trade, barter and cultural communication

Dairy products were almost never used for trade or barter. Each family processed these just for their own consumption. At the Central Asian bazars or in neighbouring countries livestock and handicrafts were traded, but no food. At the border of the steppes and the oasis – for instance in the Fergana valley, the Samarkand area and Khorezm – for a long time existed a great number of bazars specialized in selling livestock. The livestock-breeders drove their sheep, camels and horses to these bazars, that were frequented by local people interested in animals, as well as traders from abroad. In the outskirts of Samarkand such a special livestock bazar still survives, called Juma.

The most famous local animals were the 'fat tail' sheep, kept by the Uzbeks and the Tajiks, as well as the Turkmen Akhaltekyn horses and the Lokay horses, bred by the Uzbek-Lokays (Karmysheva, 1979:128). Camels used to be sold to Khiva, Bukhara, Azerbaijan, Iran and other countries. The Turkmen breeders bought camels (two humped *Baktrian*) from the Kazakhs. They crossbreed them with their dromedaries, to get powerful and hardy camels.

The breeders traded skins, leather, leather footwear, wool, woollen threads and materials, ornamented felts, specially woven carpets (which were mostly sold to the seminomadic livestock-breeders in the southern areas of Kazakhstan and Kyrgyzstan, the Turkmen oasis, the Uzbeks and the Karakalpaks). These carpets were very popular among the sedentary people of the region at the end of the 19th century; they were exported as well, to Russia and to Europe.

At the bazars the nomads used to buy agricultural products and products of city craftsmen, like cotton and silk, copper or iron pots and pans, and pottery. Pottery was expensive and nomads had to transport it in special wooden or leather cases, to prevent it from breaking.

The bazars witness the close economic and trade relationships between livestock-breeders and farmers, as well as among the various livestock-breeding peoples, with their economic specializations. For instance, in the Emirate of Bukhara, the best Kuprakhi horses (especially trained and used in the *kuprakhi*, a sports game with the carcass of a goat) could be bought from the Uzbek-Lokays, and the best rams and sheepdogs from the Uzbek-Karluks, Turks and Kungrats. With the Afghan and the Arab merchants, flour and other food products were exchanged for Afghan and Arabic wool. In spring the Afghans and the Arabs drove their rams from Afghanistan through the southern regions of Tajikistan to Kokand, where they sold them. Almost every Uzbek bought carpets and bags from the Amu Darya Turkmen, though some groups of the Uzbeks produced their own carpets (Idem, 129).

The existence of these economic connections testified that the peoples and ethnic groups of Central Asia did not live an isolated life, limited within the framework of their natural economy. The contacts, local as well as regional, secured the intercourse between different economic systems. And together with these economic relations, cultural communication took place. At the turn of the 20th century, economic relations within the region became more intense. Nomads became more and more settled and involved in agriculture; others tried to specialize in the breeding of various types of livestock. However, their traditional archaic way of life based on the patriarchal relations still prevailed. These would be more thoroughly reversed in the ensuing epoch.

Notes

1. In the Seven River region in southeastern Kazakhstan, seven rivers flow from the Djingarski Alatau to Lake Balkhash; the Ili, Karatal, Bien, Aksu, Lepsa, Baskan and Sarkand rivers. Only in winter do their flows reach the Balkhash Lake.
2. Argynbaev, 1975:197; Basilov, 1973:192-193; Peshchereva, 1927:47-48; Shaniyazov, 1973:93-94; Shaniyazov et al., 1981:178-179.
3. About the yurta's construction and design of different peoples, see: Antipina, 1962:154-174; Vostrov/Zakharov, 1989:26-38; Karmysheva, 1954:134; Mukanov, 1981:7-46; Shaniyazov, 1974:224-234; Etnografiya, 1980:27-54.

3 Modern animal husbandry in Central Asia: a call for research

Larisa Popova

Already in the 19th century, after Central Asia joined Russia, nomadic life began to change. Part of the land was given to peasants, immigrating from Russia, because of its overpopulation of farmers. As a result, the seasonal pastures, especially the winter ones, were reduced and the farmlands cut of nomadic routes and access to water reservoirs. This situation forced many breeders, especially those with only a small number of livestock, to settle down. Thus, for example, by 1914 22.4% of the Kyrghyz were leading a settled life (Tursunbaev, 1973:225).

The disaster of forced collectivization

Tsarism did not fully destroy the social and economic structure of life of the peoples of Central Asia. Even after the establishment of Soviet power in 1917, and the formation of the Soviet Union, most of the nomadic communities kept their traditional economy and way of life. The reform of 'water and land' in the middle of the twenties, mainly concerned the agricultural districts, where land was divided among the population. But the twenties and thirties would become a real tragedy for all the peasants of the country, farmers as well as livestock-breeders. The totalitarian state demanded the destruction of the class of small owners (by which the peasants were meant). The violent collectivization was accompanied by the cruel destruction of the 'middle-income' and rich owners. For ideological reasons strict settling rates for nomadic and semi-nomadic villages (*auls*) were implemented. Many administrators understood this process in a very vulgar and primitive way. In some cases it was interpreted as the forced centralization of hundreds of communities from an enormous area to one place. Others interpreted this as a directive to organize the settlements according to strict rules, with all yurtas arranged in the form of an ideal square (Abylkhozhin et al, 1989:60).

Such tactics made it impossible for the breeders to keep their herds; it resulted in the flight of livestock-breeders across the boundaries. The breeders were faced with unprecedented casualties on the animal population: in 1932 only 1.3 out of 18.5 millions of sheep were left; the number of horses reduced from 3.5 million to 885 000 in 1941; only 63 000 camels stayed alive of 1.04 million in 1935 (Idem, 63). Notwithstanding the enormous reduction of these numbers of livestock, the breeders were forced to carry out the state plan on meat and wool production. Agriculture and animal husbandry were considered an important and basic condition for the industrial development of the Soviet Union. The state enforced rigid terms of output. As a result, people were forced to shear the sheep in the severe cold, although this had to lead to the death of the animals.

This social and economical cataclysm resulted in the 1931-33 famine. Sheer need made the hungry masses concentrate in the villages and settlements near railway stations, hoping to survive. At these crowded places typhoid epidemics broke out, to which the steppe peoples were not immune. In 1920 the population of Kazakhstan counted 6.5 million people. Half of the population, that survived the famine, more then 1 million people, went to China, Mongolia, Afghanistan, Iran and Turkey. 414 000 of them would later return to Kazakhstan (Malaya Sovjetskaja Enciklopedija III, 1930; Abylkhozhin 1989:67).

Another effect of collectivization concerned the patriarchal community structure. It changed into a typical 'collective farming community' structure with remarkable characteristics of serfdom. The compelling state system resembled strongly former feudalism, but without the reciprocity principle that governed the relations between serf and landowner. People had lost traditional rights and protection (Cheshko 1990:113). In the traditional Central Asian society common law was a stable system, backed by the structure of community life. Collective farms (kolkhoz) were organized on the basis of such traditional *aul*, the community of close relatives. The collective farm brigades and sections were formed on the 'relative-neighbour' or 'tribal' principle. The head of a related group was appointed the leader of a collective farm. 'Strange' land could only be occupied after the formal and symbolic rite of selling-buying, or after going through the rite of 'gift' (Polyakov, 1989:14). Adapting these traditional relationships to Soviet needs, in some degree and in a new historical context, restored the 'patriarchal' forms of social life.

Postwar stabilization

After the stabilization of the political process, people returned to the economic basis of the former livestock breeding community: pastoralism. This extensive system is appropriate for Central Asia for the following reasons: First, because it is difficult to develop agriculture and intensive forms of animal husbandry in desert and mountain areas, due to, for instance, scarce vegetation in the desert, a severe climate and the broken ground in the mountains. Second, vast territories are only fit for pastures, offering possibilities for livestock breeding production at low costs and with minimum expenditures for equipment and labour. A third important reason is, that most inhabitants still are skilled livestock-breeders, especially the elder generation – the former nomads (Kazarevskiy, 1973:244).

These factors led to a type of pastoral livestock keeping on the state collective farms, building upon the practice and experience of former nomads and seminomads. The population, however, lives in the settlements, combining agriculture with stall breeding, processing of products, gardening, bee keeping etc. Herding is done by professional herdsmen (*chaban*), who get paid by the state (in kind or with money).

Today, collective farms on the steppe, desert and mountain areas of Central Asia combine the following main types and forms of animal husbandry:

1. Stall feeding without winter herding; when it is warm the animals are herded onto the pastures near the settlement. This form will be found in the densely populated and intensively tilled steppes and hills, with large scale cow and pig breeding kolkhoz and sovkhoz as a new trend in Kazakhstan, Kyrgyzstan and Uzbekistan.
2. Herding the livestock (like sheep, goats, milking-mares with their foals) far away from the main settlement in summertime, combined with stall feeding during the cold season. The draught animals and milk cows are stalled all year round. This form is typical for the dry steppes and mountain areas of Kazakhstan, Kyrgyzstan, Uzbekistan and Tajikistan.
3. Herding some livestock, like sheep, goats, or milking-mares with young foals, in moderate winters far from the settlement; the rest of the livestock remain close to the settlement. This form is typical for the desertlike areas of Central Asian republics (except Turkmenistan).
4. All the year round the herding of goats, sheep, milking-mares with foal on seasonal pasture lands, sometimes located far away from the main

Typical cattle-station in Kazakhstan, with a yurta near the house. In combination with brick houses yurtas are often used as kitchens or storerooms.

At the end of the day, the sheep are driven into the enclosures close to the yurta, to protect them from wild animals like foxes and wolves (Kazakhstan).

settlement (hundreds of kilometres), during at least 300 days a year. This form is typical for the desert and high mountain areas, of most of the animal husbandry sovkhoz and kolkhoz in all republics, especially Kazakhstan.

5. All year round herding the livestock at the seasonal pastures with a minimum stock of forage. This form of animal husbandry is typical in the more southern, warm and less snowy areas, especially in the deserts of Turkmenistan, Uzbekistan and south Kazakhstan, and partly for the dry high mountains of the central Tien Shan and Pamir. Here the desert sovkhoz specialize in camel and Astrakhan keeping and the sovkhoz situated high in the mountains specialized in horse and yak keeping (Kazarevskiy, 1973:254).

Livestock farms and the shepherds

The large state livestock farms, kolkhoz and sovkhoz[1], combined traditional methods of animal husbandry with modern techniques. Crucial, of course, was and is the role of the herdsmen and shepherds (*chaban*). The condition, good health and fertility of the animals depends on their knowledge of the natural conditions, and how to use the pasture lands as efficient as possible.

The work of the chaban is very risky and tough. As a rule, he is far away from his relatives. The state authorities, mostly for ideological reasons, tried to ease the chaban's daily life by constructing seasonal settlements at the pasture lands, building district centres etc. These centres keep in touch with the chaban, travelling by horse or camel, but also with cars and helicopters.[2] Many chabans also have their own field radios (Idem, 256).

Much attention was paid to the cultural and social problems of the people: medicines, mail, sometimes even the nomadic 'library' yurta are transported to the pastures. Family members of the chaban may stay with him. They are allowed to use special transportation cars. They assist in herding the livestock, as well as in processing the products. Some families only join the chaban in summer, because the children have to attend school in the settlement. But very often the children only finish primary school and start working with their chaban fathers in the pastures at a young age.

The way the chabans and their families live corresponds in some degree to the traditional culture: staying generally nomadic. The most popular house is still the yurta and not the typical standard house.[3] Moreover, in some regions people prefer a 'home-made' yurta, not a manufactured one (Konovalov, 1986:55). The wooden parts of the yurta are made by craftsmen who almost completely follow traditional technology. The fact that this portable dwelling is still in use stimulates the handicrafts and the manufacture of the decorative furniture of the yurta: the patterned mats, the woven bands for fastening the framework, the floor felts and the carpets, the different types of trunks to store, etc.[4]

Other items that are traditionally kept are, for example, the objects used for preparing meals (Idem, 74-92; Vostrov, 1956:78-80), some pottery (Abramzon, 1958:201-204), women's clothing (Konovalov, 1986:70-71) and the equipment of the saddle animals (Idem, 37-38; Zakharova, 1956:182-183). In reproducing these traditionally used objects, the livestock-breeders show the survival of traditional knowledge, connected with the peculiarities of their economic activities. The centuries-old experience and traditional skills and methods of animal husbandry are very popular and widespread even in modern practices.

Improving sheep and the year cycle

Those communities, that keep the flocks all the year round in seasonal pastures, start to drive their animals to the winter pastures in November. These winter pastures are still located in the sand hills of the deserts and the desertlike areas (Turkmenistan, Uzbekistan, Kazakhstan), or in the river valleys and ravines at the foot of the mountains (Tajikistan, Kyrgyzstan). The herds mainly consist of sheep. The local breeds, as a rule, are easy to handle, and can perfectly stand the winter herding. During the Soviet years breeding methods have been much improved. The sheep bred mainly for meat are: the Gyssar fat tail sheep (Uzbekistan, Tajikistan) (Tursunbaev, 1973:90), and the Saragene breed (Turkmenistan) (Gel'dyev et al., 1972:248). Among the breeds for wool production the most outstanding are the Tien Shan breed (Kyrgyzstan) and the Uzbek cross between the local sheep and the Lincoln ram (Sistema, 1992:91). The number of fine fleeced and half-fine fleeced sheep increases considerably. To the first type belong: the Kazakh, the Kyrghyz and the Pamir fine fleeced sheep and the Kazakh Arkharomerinos, to the second the half-fine fleeced Soviet Merinos bred in Turkmenistan (Abramzon et al., 1974:30; Alimzhanova, 1985:26). The Astrakhan fat tail sheep are of great importance in Turkmenistan, Uzbekistan, Tajikistan and are valuable for export (Shanyiazov, 1973:94-96).

Improving breeds of especially sheep, as well as rearing and breeding of totally new sorts of animals are done by many research institutes of livestock breeding, special pedigree stations and farms. The main methods of breed improvement are: selecting and combining the pairs to be crossed under perfect conditions regarding food and treatment. Artificial insemination, which is very popular and widespread is the main method of crossing the most prolific and developed animals (Kul'tura, 1967:58). The pedigree stations keep the sperm of the best rams which are frozen at a temperature of minus 193°C, and during the mating season it is delivered to the breeding farms. Before the mating starts, a ram is put with the ewes. This ram wears a special apron. For artificial insemination of about 500 sheep the sperm of one ram is sufficient (Konovalov, 1986:33).

The mating season is usually in October, so that the young are born at springtime. Bringing forth young is the hardest period of the breeding cycle. It takes place, depending on local conditions, in spring or in late winter. When the flocks are driven from the winter pastures they pass flat slopes and shallow snow. No more than ten kilometres a day can be covered. Many farms have special 'lamb-houses' or delivery rooms in the stables. These places are heated, bales with pressed lucerne and other forage are prepared, with special medicines to cure mastitis.

Skilled chabans pay special attention to the ewes; they prevent them from lying down on the cold ground, or pushing each other, in order to avoid miscarriages (Kul'tura, 1967:54). The young lambs are put in special cages in the stables, special yurtas or other buildings. The first weeks the ewes stay with their lambs; the mothers cannot abandon their little lambs. Salt is put on the lamb's back. When the mother sheep licks it, she gets used to her young more quickly and easily (Idem).

To rear young lambs, the *sakman* method is much applied. Ewes and their young are put together in groups (sakman); 10 to 15 animals at first and after a while 40 to 60 animals of the same age. The *sakman chaban* takes care of the group, herds and feeds them at the right time and in the right place (Abramzon et al., 1958:88). Each flock has five or six sakman chabans. When the lambs are two weeks old their tails are amputated (only with the fine fleeced lambs) and the rams are castrated. These operations are executed by the veterinarian or the chaban himself by the 'no blood' method. When the lambs are three to four weeks old, they join the flock.

In the districts where the Astrakhan sheep are reared, the most important activity during this period is the selection of young rams; the selected rams are brought to a central point to be killed for the astrakhan. Craftsmen from the farms come to the settlements for a preliminary preservation of the skins. These are rubbed with salt, in order to dry out. Further treatment takes place in a factory (Sistema, 1992:108).

The best young rams are selected for breeding, the others have their ears marked: a mark on the left ear shows the quality of the curl (small, middle or large), a mark on the right ear shows the quality of the breed (1st, 2nd, 3rd) (Vasil'eva, 1972:132). The colour of the skin in also very important for the breeders. Nowadays, in Turkmenistan rearing Astrakhan sheep of the colour *sur* has been developed (Lemaev, 1972:177-203).

After this period farmers have to prepare for the shearing, which starts in June or July and takes place in the spring pastures. Even in the fifties, people sheared their sheep by hand with simple shears. Now the shearing process is fully mechanized, although some people still prefer the traditional way of sheering in spring (Vasil'eva, 1972:133). The sheep are sheared with electric shears on wooden tables. Since mechanization eased this work, women also participate (in the past it was a fully men's job). Nonetheless, shearing needs great skill. The farms prefer craftsmen who shear fifty to sixty sheep and rams a day (with common scissors this is not more then ten to twelve). One sheep gives about 2 kg of wool (Konovalov, 1986:35); some breeds give 3 to 3.5 kg of wool (and even more) a year (Kazakhskiy, 1971:5-6).

The sheared wool is delivered to a tally desk, classified and then pressed in bales and brought to the storehouse. After being sheared the sheep are bathed in a special solution to prevent them from mange (Kul'tura, 1967:60).

After shearing, the flocks are driven to the summer pastures, which are usually located in the high mountain zone, on the Alpine fields (Kyrgyzstan, Tajikistan), as close as possible to the wells (Turkmenistan), or in the forest steppes of northern Kazakhstan. In the cool areas the cattle is herded in daytime – from six a.m. until ten p.m. The flocks are driven before the sun, so that the eyes cannot be blinded. In the hot areas the sheep rest during the daytime, and at night-time when it is cooler, they are herded. In summer the sheep are watered two times a day.

In September the young lambs are separated from their mothers and form new groups, according to their sex, weight and breed. From the sheep that will be driven to the winter pastures, the weak ones are selected and left at special farms, close to the central farmsteads, in the hands of experienced breeders (Idem). In September the flocks are herded to the pastures quite close to the settlements. In October the sheep are sheared for the second time, but specialists recommend the shearing of fine fleeced sheep only once a year, in spring (Konovalov, 1986:35). After that follows the mating campaign, and when this is finished, people have to start to prepare for the harsh winter herding.

In summer and fall, the people prepare a supply of winter forage: hay, grass flour, pressed up grass etc. For this reason special temporary brigades of mowers, and people who lay in the forage, are organized. Most of the arable land is used for high yield crops. The climatic conditions give good crops of lucerne, clover and other grasses, four to five times a year. On the hay-making farms production capacity is even higher (600 to 700 kg per hectare).

Goats. Many farms rear goats, providing very good wool and down. The Soviet 'woollen' breed provides the most valuable wool called *mocker*. A she-goat gives 1.5 to 2 kg of wool a year and a he-goat 2.5 to 3 kg. The breed 'black down' provides 0.4 to 0.5 kg of down a year; the big goats give 0.7 to 0.8 kg. A flock numbers 300 or 400 animals at the most.

Shearing of the goats starts in April. It is very important to be ahead of the moulting season, for then the wool quality deteriorates. From goat skin the best leather is made, like morocco, kid leather, chamois (Sistema, 1992:109).

Horses. Horses of livestock-breeders are at the pastures all the year round. In April they are driven to the spring pastures, where they bring forth their 'younger generation'. When the foals are stronger, the herds are driven to the summer pastures, usually at the beginning of June. At these pastures *koumiss*, the fermented mare's milk, is made by the women, who travel to the pastures especially for that reason. Before the mares are milked, they let the foal suck. When the women take over and milk the mare, the foal is kept near his mother. After the fifth or the sixth milking, the foals again are set free (Kul'tura, 1967:64).

The horse herds are formed in summer, strictly according to age and sex. The 'mare' herd consists of about 100 to 150 mares, the 'young' herd of 250 to 300 horses and the herd of stud-horses of 20 to 25 horses (Sistema, 1992:126). For the mating special herds are formed: 10 to 15 (20 to 25) mares and one stallion.

In Uzbekistan, Karábine horses are preferred; the Kyrghyz and the Lokay breeds are preferred at Pamir; the Iomud and the Akhal-Teke breeds in Turkmenistan and the Kazakh people love the horses of the Yakut, the Bashkir and the Kazakh *jabe* breeds. The most famous have always been the real Akhal-Teke stallions. However, during the Soviet period, Turkmen horse breeding seriously declined. Nowadays the conditions in the state stables of Ashkhabad and the Mariyskaya district are very bad: there is not enough forage base and no possibility to herd the horses (Yanborisov, 1987:35-36).

Camels. Camel breeding is traditional for all peoples of Central Asia. The camel is the very best beast of burden even nowadays; he moves perfectly over the sands and climbs mountains where a car cannot get through. The baktrian camel is popular in the northern areas of the region and in the desert areas of Turkmenistan; other republics prefer the dromedary.

Camels are reared in herds. To have them mate, they are put in special groups consisting of three year old females and four to five year old males. The dromedary's pregnancy lasts 385 to 389 days and that of the baktrian 413 to 417 days. Taking care of the young camels is a hard and labour-intensive job. In preparation, felt blankets are made and buildings heated, because the young camels can easily catch a cold. Those buildings contain warm covers, hay, salt (which camels need a lot) and water of an appropriate temperature. Each little camel gets a half litre of vegetable oil, to clean its digestive system. Special attention is given to the problem that young camels are predisposed to various infectious and non-infectious diseases (Sistema, 1992:132, 157-159).

Camels always need to be fed with fresh green fodder. Therefore, spring herding is of great importance, for in this period green vegetation grows in the desert sands (Fedorovich, 1973:213). The camels always reach out for the most juicy and green parts of the plants in fall and winter. When the snow starts, the chabans herd their camels only to pastures without snow, or with high vegetation, for camels cannot get grass from under the snow. All year round camels need a lot of additional salt. It is put in a special nose-bag, so that each animal gets its 100 grams of salt a day.

The camels are sheared once a year, in spring. For this purpose a big knife is used. The very best wool is given by young camels. There are two types of wool: the long hair type *shudah* and the curly *jabagyh*. One camel provides about 5 kg of wool; a dromedary two times less (Kul'tura, 1967:70). Apart from the wool, people highly value camel's milk, of which they make a very healthy drink, called *shubat*, as well as butter and cheese. A camel gives about 300 to 350 kg of meat, but the taste is not very nice (Sistema, 1992:134).

Training camels for riding starts as soon as they are two years old. They get a hole in the nose with a ring (*murunduk*) through it. The pack-camel carries a weight half its own, covering 30 to 35 kilometres a day, at a rate of 4.5 to 5 km/hour (Idem, 135).

Horned cattle. The eastern Pamir Kyrghyz breed yaks, which give meat, milk and wool. Nowadays in the Murgaba area of the far eastern part of Tajikistan are special yak breeding farms (Shibaeva, 1973:117). Yaks use pastures which other animals cannot reach. The Pamir have more than one million hectares of such hardly accessible land. The best forage for yaks is *teresken*, growing 1 000 metres above sea level.

Yaks are herded for 2 to 2.5 months to the high pastures, then they are gradually driven to the valleys. Each herd of yaks consists of 80 to 100 adult animals, 110 to 120 young animals and 8 to 10 yak bulls. Each herd has two or three breeders/chaban. A he-yak weights about 400 to 520 kg and a she-yak 240 to 350 kg. After slaughtering, about 50 to 53% of the meat is left over. A she-yak gives about 600 litres of milk a year (Zhivotnovodstvo, 1985:69-70).

Most of the horned cattle are reared in farms with stalls, with 'pasture camps' and 'stall pasture' keeping of cattle. The animals are not fit for year-round herding. The former Soviet republics know a considerable variety of cattle breeds. The milk breeds are: the black-patched (Holstein, Dutch), the red-steppe (the Estonian Red, the Latvian Brown (reddish-brown)), the Bushuev, the Aulie-Ata, the Swiss milking type etc. Meat cows vary between: Santa-Gertruda, the Kazakh Whiteheaded, the Kalmyk, and the Aberdyn-Angus (Sistema, 1992:41-46).

Work at the farms is mostly mechanized: there are automatic watering systems, machines to provide the cattle with forage, milking machines, and machines to clean the enclosures. At the farms most of the work is done by the women. In summer the men (herdsmen) herd the cows to the nearer pastures. In the fifties the dairy farms even moved to the mountains (Kyrgyzstan) to herd their cattle on the juicy fresh grass (Abramzon et al., 1958:95).

New and non-traditional kinds of animal husbandry (pigs and poultry), became widespread in Central Asia during the Soviet years. In the thirties they were introduced by force, by the government. As eating pork is condemned by Islam, only western immigrants work at the pig breeding farms, mostly Russians and Ukrainians (Idem, 99).

A future for private animal husbandry?

This listing of modern Central Asian animal husbandry concerned the state livestock farms: kolkhoz and sovkhoz. In the seventies and eighties, restrictions on establishing private farms were weakened, and recently disappeared. Generally speaking, private farming has since become the main source of income for the majority of the population. State salaries are too modest to support one's family – traditionally with many children – whereas festivities and festivals are also very expensive. People tend to spend large amounts of money on marriages, funerals, childbirth etc.

Peasants face a hard life and they are urged to realize a maximum output from their own land. The fact that they almost do not pay taxes, and that they heavily rely on the unpaid labour of their children and other members of family, helps the farm to survive. Private income is usually spent on livestock, and one does not need to be surprised that, in the eighties an increase in the less expensive, traditional methods could be noticed in the livestock breeding areas (Polyakov, 1989:35).

Sheep are a means of increasing wealth and their number is augmenting in every private farm. S.P. Polyakov gives the following data: in the middle of the eighties, in the sovkhoz of M.V. Frunze (in the Osh district of Kyrgyzstan) the state flock numbered 4 500 sheep, whereas 20 000 were privately owned. The same accounts for the horned cattle. This tendency also indicates a certain accumulation of wealth among certain families (1989:24).

However, such an uncontrolled increase of privately owned animals could harm nature, for they are herded at random. As a result, the pastures deteriorate, the mountain forests are ruined, and thus the hydrology of the area changes for the worst. The sheep and goats also destroy certain kinds of vegetation. Sheep eat the 'archy' bushes that grow in the mountains, causing more and more avalanches (harming both nature and people) (1989:36).

Researchers know little of modern private livestock farming.[5] During the Soviet period, from the thirties to the sixties, anthropologists did not explore this subject at all, since few private farms existed. But nowadays this situation has totally changed. The number of private farms keeps growing, whereas the state farms collapse. Private animal husbandry in Central Asia, has a crucial influence on the process of economic and social change in the Central Asian republics. It has therefore become a subject, which really needs scientific attention from the republican governments, animal husbandry specialists and others involved.

Sale of a sheep to a private buyer in the bazaar of Almaty.

Main street in the Kazakh sheep breeding kolchoz which is named after the sixty years old Soviet Union. The kolchoz dates from 1978.

Notes

1. See note 4 chapter 1.
2. Because of the severe economical crisis at the beginning of the nineties in the former Soviet republics, the use of expensive means of transportation was considerably reduced.
3. Not only do the chabans live in yurtas, but also people in the settlements. In the hot period they put the yurta on the farmyard and live in it, and not in the house (Abramzon et al., 1958:172). The yurta is still an important ritual place to celebrate weddings, to receive guests.
4. Some decorative elements of the yurta can be found in 'normal' houses (Zakharova, 1956:156-162).
5. There are only a few works on private animal husbandry of the mountain Tajiks and the peoples living at the foot of the Pamir (Monogarova, 1972:68-70; Nemenova, 1963:73; Rakhimov, 1963:48; Rakhimov (b), 1963:67-58).

Bibliography

Abramson, C.M., *Kirgizi i ikh etnogeneticheskie i istoriko-kul'turnye svyazi* (The Kyrghyz and their ethnogenetic, cultural and historical relations). Leningrad, 1971.

Abramzon, S.M., Antipina, K.I., Vasil'eva, G.P. et al., 'Byt kolkhoznikov kirgizskikh seleniy Darkhan i Chichkan'. *Trudy Instituta etnografii. Novaya seriya*, t. XXXVII ('The way of life of the kolkhoz farmers in the Kyrghyz settlements of Darkhan and Chychkan'. Works of the Ethnographic Institute. New Series, Vol. XXXVIII). Moskva, 1958.

Abramzon, S.M., Simakov, G.N., Firshteyn, L.A., 'Nov' kirgizskogo sela'. *Sovetskaya etnografiya*, nr. 5 ('The present-day Kyrghyz settlement'. Soviet Ethnography, nr. 5). 1974.

Abylkhozhin, Zh.B., Kozybaev, M.K., Tatimov, M.B., 'Kazakhstanskaya tragediya'. *Voprosy istorii*, nr. 7 ('The tragedy of Kazakhstan'. Historical Questions, nr. 7). 1989.

Alimzhanova, L.V., *Budu zootekhnikom* (I will be a zoo specialist). Alma-Ata, 1985.

Allworth, Edward (ed.), *The nationality question in Soviet Central Asia*. New York, 1973.

Andreev, M.S., *Tadzhiki doliny Khuf* (The Tajiks of the Khuf valley). Stalinabad, 1958.

Antipina, K.I., *Osobennosti material'noy kul'tury i prikladnogo iskusstva yuzhnykh kirgizov* (Peculiarities of the culture and applied art of the south Kyrghyz). Frunze, 1962.

Argynbaev, Kh., 'Narodnye obychai i pover'ya kazakhov, svyazannye so skotovodstvom'. *Khozyaystvenno-kul'turnye traditsii narodov Sredney Azii i Kazakhstana* ('Popular beliefs and customs of the Kazakhs, connected with livestock breeding'. The cultural and economical traditions of the peoples of Middle Asia and Kazakhstan). Moskva, 1975.

Avdakushin, I.S., 'Sanitarnyy obzor Amudar'inskogo otdela c 1887 po 1891 g.'. *Sbornik materialov dlya statistiki Syrdar'inskoy oblasti* ('A sanitary review of the Amu Darya Department from 1887 till 1891'. Statistical material of the Syr Darya District). Tashkent, 1892.

Bartol'd, V.V., *Sovremennoye sostoyanie i bluzhayshie zadachi izucheniya turetskikh narodnostey* (Present-day state and tasks of the study of the Turkish peoples). Moskva, 1968.

Basilov, V.N., 'Khozyaystvo zapadnykh turkmen-yomudov v dorevolyutsionnyy period i svyazannye s nim obryady i verovaniya'. *Ocherki po istorii khozyaystva narodov Sredney Azii i Kazakhstan* ('The economy of the pre-Revolutionary period, rites and beliefs connected with it'. Essays on the history of the economy of the peoples of Middle Asia and Kazakhstan). Leningrad, 1973.

Bayalieva, T.D., *Doislamskie verovaniya i ikh perezhitki u kirgizov* (Pro-Islam religions and their survivals with the Kyrghyz). Frunze, 1972.

Bennigsen, A. and S. Enders Wimbush, *Muslims of the Soviet Empire, a guide*. London, 1985.

Buschkow, Walentin, *Tadschikistan vor dem Bürgerkrieg. Eine traditionelle Gesellschaft in der Krise*. Köln, 1993. (Berichte des Bundesinstituts für ostwissenschaftliche Studien, Nr. 26)

Carrère d'Encausse, H., *Islam and the Russian Empire*. London, 1988.

Castagné, J., 'Magie et exorcisme chez les Kazak-Kirghizes et autres peuples turcs orientaux'. *Revue des Etudes Islamiques* IV, 1930, p. 53-151.

Cheshko, S.V., 'Srednyaya Aziya i Kazakhstan: sovremenoye sostoyanie i perspektivy natsional'nogo razvitiya'. *Rasy i narody*, t. 20 ('Middle Asia and Kazakhstan: The present-day situation and perspectives of national development'. Races and peoples, Vol. 20). 1990.

Clem, R.S., 'The impact of demographic and socioeconomic forces upon the nationality question in Central Asia'. In: Allworth, E. (ed.), *The nationality question in Soviet Central Asia*. New York, 1973.

Czaplicka, M.A., *The Turks of Central Asia, in history and at the present-day*. Amsterdam, 1973.

Dachšlejger, G.F., 'Sesshaftwerdung von Nomaden. Erfahrungen über die Dynamik traditioneller sozialer Einrichtungen (am Beispiel des kasachischen Volkes)'. In: *Die Nomaden in Geschichte und gegenwart (Beiträge zu einem internationalen Nomadismus-Symposium am 11. und 12. Dezember 1975 im Museum für Völkerkunde Leipzig)*. Berlin, 1981. (Veröffentlichungen des Museums für Völkerkunde zu Leipzig, Heft 33)

Eliade, M., *Shamanismus und archaische Extasetechnik*. Zürich/Stuttgart, n.d.

Etnografiya karakalpakov. XIX-nachalo XX veka (The ethnography of the Karakalpak. The 19th-beginning of the 20th century). Tashkent, 1980.

Fedorovich, B.A., 'Prirodnye usloviya aridnykh zon SSSR i puti razvitiya v nem zhivotnovodstvo'. *Ocherki po istorii khozyaystva narodov Sredney Azii i Kazakhstana. Trudy Instituta etnografii AN SSSR. Novaya seriya*, t. XCVIII ('The natural conditions of Asiatic USSR and the development of animal husbandry in this area'. Essays on the economical history of the peoples of Middle Asia and Kazakhstan. Works of the Ethnographic Institute of the SSSR. New Series, Vol. XCVIII). Leningrad, 1973.

Gatenby, R.M., *Sheep*. The Tropical Agriculturalist. London, 1991.

Gel'dyev, K., Logvinov, L, Gushliev, O., 'Uluchenie plemennykh i produktivnykh kachestv saradzhinskikh ovets'. *Kormovaya baza i zhivotnovodstvo Turkmenistana* ('The improvement of pedigree and productive qualities of the Saradjin sheep'. Food and livestock breeding in Turkmenistan). Ashkhabad, 1972.

Gressler, S., *Kasachstans schwieriger Weg in die Unabhängigkeit. Ein Erfahrungsbericht*. Köln, 1993. (Berichte des Bundesinstituts für ostwissenschaftliche und internationale Studien, Nr. 12)

Grobe-Hagel, K., 'Islamische Atommacht, Kasachstan ist der bevorzugte Partner Russlands'. *Der Ueberblick* 2, Juni 1992.

Halbach, U., *Islam, Nation und politische Oeffentlichkeit in den zentralasiatischen (Unions-)Republiken*. Köln, 1991. (Berichte des Bundesinstituts für ostwissenschaftliche und internationale Studien, Nr. 57)

Hayit, B., *Sovjetrussische Orientpolitik am Beispiel Turkestans*. Köln/Berlin, 1962.

Jettmar, K., *Art of the steppes*. London, 1964.

Karmysheva, B.Kh., 'O torgovle v vostochnykh bekstvakh Bukharskogo khanstva v nachale XX veka v svyazi s khozyaystvennoy spetsializatsiey'. *Tovarno-denezhnye otnosheniya na Blizhnem i Srednem Vostoke v epokhy srednevekov'ya* ('On trade in the eastern parts of the Kingdom of Bukhara in the beginning of the 20th century'. Trade money in the Medieval Near and Middle East). Moskva, 1979.

Karmysheva, B.Kh., 'Tipy skotovodstva v yuzhnykh rayonakh Uzbekistana i Tadzhikistana, konets XIX-nachalo XX veka'. *Sovetskaya etnografiya*, nr. 3 ('The types of animal husbandry in the southern regions of Uzbekistan and Tajikistan, end 19th-beginning 20th century'. The Soviet Ethnography, nr. 3). 1969.

Karmysheva, B.Kh., 'Uzbeki-lokaytsy Yuzhnogo Tadzhikistana'. *Trudy Instituta istorii, arkheologii i etnografii AN Tadzhikskoy SSR,* t. XXVIII ('The Uzbek-Lokay of south Tajikistan'. Works of the Institute of Ethnography, History and Archeology of the Tajik Republic, Vol. XXVIII). 1954.

Kaul'bars, A.V., 'Nizov'ya Amudar'i'. *Zapiski Russkogo geograficheskogo obshchestva,* t. IX ('The lower reaches of the Amu Darya'. Works of the Russian Geographic Society, Vol. IX). Cankt-Peterburg, 1881.

Kazakhskiy epos, t. 1 (The Kazakh Epos, Vol. 1). Alma-Ata, 1963.

Kazakhskiy nauchno-issledovatel'niy institut zhivotnovodstvo (The Kazakh Scientific Research Institute of Animal Husbandry). Alma-Ata, 1971.

Kazarevskiy, O.R., 'Sovremennye formy pastbishchnogo zhibotnovodstva v pustynnykh i gornykh rayonakh Kazakhstana i respublik Sredney Azii'. *Ocherki po istorii khozyaystva narodov Sredney Azii i Kazakhstana. Trudy instituta etnografii AN SSSR. Novaya Seriya,* t. XCVIII ('Modern forms of pasturage livestock breeding in the desert and mountain areas of Middle Asia and Kazakhstan'. Essays on the economical history of the peoples of Middle Asia and Kazakhstan. Works of the Ethnographic Institute of the SSSR. New Series, Vol. XCVIII). Leningrad, 1973.

König, V., 'Skotovodcheskoe khozyaystvo u tekintsev Akhala vo vtoroy polovine XX veka'. *Trudy Instituta istorii, arkheologii i etnografii AN Tukmenskoy SSR,* t. VI ('The animal husbandry economy of the Akhal-Tekyn in the second part of the 19th century'. Works of the Institute of Ethnography, History and Archaeology of the Turkmen Republic, Vol. VI). 1962.

Khozyaystvo kazakhov na rubezhe XIX-XX vekov (The Kazakh economy on the line of 19th-20th century). Alma-Ata, 1980.

Konovalov, A.V., *Kazakhi Yuzhnogo Altaya* (The Kazakhs of the south Altay). Alma-Ata, 1986.

Krist, G., *Alone through the forbidden land: journeys in disguise through Soviet Central Asia.* London, 1992. (Translation of the 1937 German edition)

Kul'tura i byt kazakhskogo kolkhoznogo aula (The culture and way of life of the Kazakh kolkhoz aul). Alma-Ata, 1967.

Lemaev, K.E., 'Opyt sozdaniya stada karakul'skikh ovets sur v plemzavode "Talimardzhan"'. *Kormovaya baza i zhivotnovodstvo Turkmenistana* ('The experience to create a flock of Astrakhan Sur sheep in the breeding farm "Talimardjan"'. Food and animal husbandry in Turkmenistan). Ashkhabad, 1972.

MacLean, F., *Back to Bokhara.* London, 1959.

Monogarova, L.F., *Preobrazovaniya v bytu i kul'ture pripamirskikh narodnostey* (Changes in way of life and culture with the Pripamir peoples). Moskva, 1972.

Mukanov, M.S., *Kazakhskaya yurta* (The Kazakh yurta). Alma-Ata, 1981.

Nemenova, R.L., 'Tadzhiki Varzoba'. *Izvestiya AN Tadzhikstoy SSR. Otdelenie obshchestvennykh nauk,* nr. 1 (32) ('The Varzobah Tajiks'. The news of the Republic of Tajikistan. The Social Science Department, nr. 1 (32)). Dushanbe, 1963.

Obzor Zakaspiyskoy oblasti s 1882 po 1890 g., izd. 2 (Review of the Zacaspian District from 1882 till 1890, 2nd ed.). Askhabad, 1897.

Orazov, A., 'Nekotorye voprosy skotovodcheskogo khozyaystva Severo-Zapadnoy Turkmenii v kontse XIX-nachale XX veka'. *Trudy Instituta istorii, arkheologii i etnografii AN Turkmenskoy SSR*, t. VI ('Some questions of the animal husbandry economy of north and east Turkmenistan at the end of the 19th-beginning of the 20th century'. Works of the Institute of Ethnography, History and Archaeology of the Turkmen Republic, Vol. VI). 1962.

Ovezberdyev, K., 'Materialy po etnografii turkmen-sarykov Pendinskogo oazica'. *Trudy Instituta istorii, arkheologii i etnografii AN Turkmenskoy SSR*, t. VI ('Material on the ethnography of the Turkmen-Saryk of the Pendyn Oasis'. Works of the Institute of Ethnography, History and Archaeology of the Turkmen Republic, Vol. VI). 1962.

Oxus. 2000 Jahre Kunst am Oxus-Fluss in Mittelasien. Neue Funde aus der Sowjetrepublik Tadschikistan. Zürich, 1989.

Pellat, Ch., 'Ibil'. In: *Encyclopaedia of Islam* III, 1971, p. 665 et seq.

Peshchereva, E., 'Molochnoye khozyaystvo gornykh tadzhikov i nekotorye cvyazannye s nim obychai'. Andreev, M.S., *Po Tadzhikistanu* ('The milking industry of the mountain Tajiks and some customs connected with it'. Andreev, M.S., On Tajikistan). Tashkent, 1927.

Polyakov, S.P., *Traditsionalizm v sovremennom srednya-aziatskom obshchestve* (Traditionalism in modern Middle Asiatic society). Moskva, 1989.

Potanov, O.P., 'Osobennosti material'noy kul'tury kazakhov'. *Sbornik Muzeya antropologii i etnografii AN SSSR*, t. 12 ('Peculiarities of the Kazakh culture'. Works of the Museum of Anthropology and Ethnography of the USSR, Vol. 12). 1949.

Rakhimov, M., 'Nekotorye rezul'taty raboty Garmskoy etnograficheskoy ekspeditsii 1954 goda'. *Izvestiya AN Tadzhikstoy SSR. Otdelenie obshchestvennykh nauk*, nr. 1 (32) (Some results of the 1954 Garmsky ethnographical expedition'. The News of the Republic of Tajikistan. The Social Science Department, nr. 1 (32)). Dushanbe, 1963.

Rakhimov, M., 'Zeravshanskaya etnograficheskaya ekspeditsiya 1958-1961 godov'. *Izvestiya AN Tadzhikstoy SSR. Otdelenie obshchestvennykh nauk*, nr. 1 (32) ('The Zeravshansky ethnographical expedition of 1958-1961'. The News of the Republic of Tajikistan. The Social Science Department, nr. 1 (32)). Dushanbe, 1963.

Het Rijk der Scythen. Zwolle, 1993/4.

Rosenbohm, A., *Halluzinogene Drogen im Shamanismus. Mythos und Ritual im kulturellen Vergleich*. Berlin, 1991.

Rossetti, B., *Die Turkmenen und ihre Teppiche. Eine ethnologische Studie*. Berlin, 1992.

Rypka, J., *History of Iranian literature*. Dordrecht, 1968.

Schletzer, D. und R., *Alter Silberschmuck der Turkmenen*. Berlin, 1983.

Gronbech, K., 'The Turkish system of kinship'. In: *Studia Orientalia, Ioanni Pedersen... dicata*. Copenhagen, 1953, pp. 124-129.

Shaniyazov, K., *K etnicheskoy istorii uzbekskogo naroda* (The ethnic history of the Uzbek people). Tashkent, 1974.

Shaniyazov, K., 'Otgonnoye skotovodstvo u uzbekov'. *Ocherki po istorii khozyaystva narodov Sredney Azii i Kazakhstana. Trudy Instituta etnografii AN SSSR. Novaya seriya*, t. XCVIII ('The Uzbek pasture livestock breeding'. Essays on the economical history of the peoples of Middle Asia and Kazakhstan. Works of the Ethnographic Institute of the SSSR. New Series, Vol. XCVIII). Leningrad, 1973.

Shaniyazov, K., 'Otgonnoe zhivotnovodstvo u uzbekov'. *Ocherki po istorii khozyaystva narodov Sredney Azii i Kazakhstan* ('The Uzbek livestock breeding'. Essays on the economical history of the peoples of Middle Asia and Kazakhstan). Leningrad, 1973.

Shaniyazov, K., *Uzbeki-karlyki* (The Uzbek-Karlyk). Tashkent, 1964.

Shaniyazov, K. and Ismailov, Kh.I., *Etnograficheskie ocherki material'noy kul'tury uzbekov kontsa XIX-nachala XX veka* (Ethnographic essays on the culture of the Uzbeks at the end of the 19th-beginning of the 20th century). Tashkent, 1981.

Shibaeva, Yu.A., 'Zhivotnovodstvo u kirgizov Vostochnogo Pamira'. *Ocherki po istorii khozyaystva narodov Sredney Azii i Kazakhstana. Trudy Instituta etnografii AN SSSR. Novaya seriya*, t. XCVIII ('Kyrghyz livestock breeding in the eastern Pamir'. Essays on the economical history of the peoples of Middle Asia and Kazakhstan. Works of the Ethnographic Institute of the SSSR. New Series, Vol. XCVIII). Leningrad, 1973.

Sistema vedeniya zhivotnovodstva v Uzbekistana (The organization of the animal husbandry system in Uzbekistan). Tashkent, 1992.

Snesarev, G.P., *Relikty domusul'manskikh verovaniy i obryadov u uzbekov Khorezma* (The relics of the pre-Moslem beliefs and the rites of the Khorezm Uzbeks). Moskva, 1969.

Stadelbauer, J., *Bahnbau und kulturgeographischer Wandel in Turkmenien*. Berlin, 1973.

Tursunbaev, A., 'Perekhod k osedlosti kochevnikov i polukochevnikov Sredney Azii i Kazakhstana'. *Ocherki po istorii khozyaystva narodov Sredney Azii i Kazakhstana. Trudy Instituta etnografii AN SSSR. Novaya seriya*, t. XCVIII ('The transition to a settled life of the Middle Asiatic and Kazakh nomads and half-nomads'. Essays on the economical history of the peoples of Middle Asia and Kazakhstan. Works of the Ethnographic Institute of the SSSR. New Series, Vol. XCVIII). Leningrad, 1973.

Vasil'eva, G.P., 'Materialy po zhilishchu turkmen Murgabskogo oazisa'. *Polevye issledovaniya Instituta etnografii* ('Material on the Turkmen of the Murgaba oasis'. Field researches of the Ethnographic Institute). Moskva, 1983.

Vasil'eva, G.P., *Preobrazovaniya byta i etnicheskie protsessy v severnom Turkmenistana* (The changing of daily life and the ethnic processes in north Turkmenistan). Moskva, 1972.

Vasil'eva, G.P., 'Turkmeny-nokhurli'. *Trudy Instituta etnografii AN SSSR*, t. XXI ('The Turkmen-Nokhurly'. Works of the Ethnographic Institute of the USSR, Vol. XXI). 1954.

Vostrov, V.V., 'Kazakhi Dzhanybekskogo rayone Zapadno-Kazakhstanskoy oblasti'. *Trudy Instituta istorii, arkheologii i etnografii*, t. 3 *Etnografiya* ('The Kazakhs of the Janibek area of west Kazakhstan'. Works of the Institute of History, Archaeology and Ethnography, Vol. 3. Ethnography). Alma-Ata, 1956.

Vostrov, V.V. and Zakharov, I.V., *Kazakhskoye narodnoe zhilishche* (The Kazakh popular house). Alma-Ata, 1989.

Yanborisov, V.R., 'Perspektivy razvitiya turkmenskogo konevodstvo'. *Kul'tura etnosa i etnicheskaya istoriya* ('The perspectives of Turkmen horse breeding'. Ethos culture and ethic history). Leningrad, 1987.

Zakharova, I.V., 'Material'naya kul'turf kazakhov-kolkhoznikov Yugo-Vostochnogo Kazakhstana'. *Trudy Instituta istorii, arkheologii i etnografii*, t. 3, *Etnografiya* ('The material culture of the Kazakh kolkhoz farmers in southeast Kazakhstan'. Works of the Institute of History, Archaeology and Ethnography, Vol. 3, Ethnography). Alma-Ata, 1956.

Zhivotnovodstvo Tadzhikistana (Animal husbandry in Tajikistan). Dushanbe, 1985

CIP-DATA KONINKLIJKE BIBLIOTHEEK,
THE HAGUE

Leeuwen, Carel van

Nomads in Central Asia : animal husbandry and culture in transition (19th-20th century) / [authors:] Carel van Leeuwen, Tatjana Emeljanenko and Larisa Popova ; [photogr.: Jaap de Jonge ... et al. ; cartogr.: K. Prins ... et al.]. - Amsterdam : Royal Tropical Institute. - Ill., maps, photo's
Publ. accompanies the exhibition "Nomaden in Centraal-Azië", on show in the Tropenmuseum in Amsterdam from 1 December 1994 to 1 August 1995.
- With ref.
ISBN 90-6832-252-4
NUGI 653/835
Subject headings: nomads ; Central Asia / animal husbandry ; Central Asia

Editing chapters 2 and 3: Gertruud Alleman - Amsterdam
Photographs: Jaap de Jonge, Irene de Groot, Carel van Leeuwen - KIT, Amsterdam; Russian Ethnographical Museum - St. Petersburg; Central State Museum of Kazakhstan - Almaty
Cartography: K. Prins, M. Rieff Jr
Cover and graphic design: Nel Punt - Amsterdam
Printing: Veenman drukkers - Wageningen

ISBN 90 6832 252 4
NUGI 653/835